The TOTALLY USELESS HISTORY OF SCIENCE

PRIMVM MOBILE

CRISTALLINE

FIRMAMENT

FIER
AER
YEARTH

CŒLIFER ATLAS

Hic canet errantê Lunam, Solisq; labores
Arcturūq;, pluuiasq; hyad.gëinosq; triões

ID.

The
TOTALLY USELESS
HISTORY
OF
SCIENCE

IAN CROFTON

Quercus

Contents

6

Introduction

The purpose of the present volume is not so much to instruct and enlighten, as to wander idly along the less-travelled byways of scientific history. Down these dusty lanes one comes across all kinds of colourful characters – eccentrics, monsters, quacks, hoaxers and frauds – not to mention a plethora of mad experiments, astonishing anticipations of future developments, crazed ideas and wild speculations.

The annals of science are full of theories that owe more to wishful thinking, false reasoning, bigotry, prejudice and pure gullibility than empirical evidence. The ancient Greek philosopher Anaximander, for example, held that the first men and women must have emerged whole at puberty, bursting out of fish-like creatures that had themselves been formed from warmed-up water and mud. A millennium later, Isidore of Seville was insisting that contact with menstrual blood would make crops fail and dogs go mad. Even at the dawn of the Scientific Revolution, people clung to old beliefs. In 1555, for example, the Archbishop of Uppsala in Sweden reported that swallows spend their winters at the bottom of the northern lakes, while in the 1630s Scipio Chiaramonti, professor of mathematics and philosophy at the University of Pisa, refuted Galileo by pointing out that the Earth could not possibly move because, unlike animals, it has neither limbs nor muscles.

Scientists in more recent times have not been immune from allowing faith to overwhelm reason. In the 19th century, for example, a noted Astronomer Royal for Scotland rejected the metric system on the grounds that the inch was a unit of measurement ordained by God, while in America a physician claimed that he had been divinely guided to conclude that pi equals exactly 3.2. A bill to this effect was introduced into the state legislature of Indiana.

Some characters are not so much deluded dupes as wilful pedlars of delusion. Heading the list are the unscrupulous concocters of universal cure-alls. The main selling point, more often than not, of such tonics is that they will restore youthful vigour, 'replenish the crispy fibres' and reverse 'impotency and seminal weakness'. With the dawn of the 20th century, the quacks lighted upon a new health-giving quality to peddle: radioactivity. Hence Radithor – a 'cure for the living dead' that contained radium 226 and 228 and thus sent many to a premature grave (in lead-lined coffins).

Some hoaxers, however, are no more than harmless pranksters. There is a mischievous tradition in which distinguished physicians submit to learned medical journals the occasional case history that is not entirely genuine. An early example is a case from the American Civil War reported in the *American Medical Weekly* in 1874, in which a bullet took off the testicle of a Confederate soldier and continued on its way into a Southern belle, who was thus (so the author would have his readers believe) inseminated.

Truth can be stranger than fiction. Certain experiments, for example, sound so bizarre one would think that someone was pulling one's leg. From the early years of the Scientific Revolution one might mention Sanctorius of Padua meticulously weighing his own excrement, Richard Lower transfusing the blood of a sheep into a man, and the alchemist Hennig Brand extracting a new element from vats of stale urine. The 18th century saw a growth in popular demonstrations of the wondrous effects of electricity: one impresario drew sparks out of the nostrils of a boy suspended in the air by silken cords; another lined up monks in a circuit and watched them all jump at once as he applied a charge; a third drew large audiences keen to see what would happen when an electrical rod was inserted into the rectum of a recently executed criminal.

A number of experimenters have taken a stronger ethical line, using themselves as guinea pigs before trying out their theories on others. The 18th-century surgeon John Hunter deliberately infected himself with 'venereal matter' to see whether syphilis and gonorrhoea were the same disease, while in the following century Dr Nicholas Chervin ate the 'bloody black vomit' of yellow-fever victims to show that the disease was not transmitted by human contact. Perhaps most heroic of all was a certain Dr Hildebrandt, who at the end of the 19th century tested the efficacy of spinal anaesthesia by allowing his colleague to burn him, stab his thigh, squeeze his testicles and hit his shins with a hammer – so demonstrating beyond doubt that he could feel nothing below the waist.

Some figures have been ahead of their times, anticipating future developments by many centuries. Hero of Alexandria built a steam engine in the 1st century AD, Abbas Ibn Firnas attempted to fly in 875, Cyrano de Bergerac described what sounds like a ramjet in 1637, and, at the end of the 18th century, the Reverend John Michell outlined the concept of black holes. Others have been distinctly *behind* their times, conforming to Arthur C. Clarke's First Law: 'When a distinguished but elderly scientist . . . states that something is impossible, he is very probably wrong.' Thus in 1900, five years before Einstein's special theory of relativity, Lord Kelvin declared that there was nothing new to be discovered in physics – having previously discounted the possibilities of flight in a heavier-than-air machine, and firmly opined that there was no future in radio. To his eternal embarrassment, in 1957 Sir Harold Spencer Jones, the Astronomer Royal, loftily declared that 'Space travel is bunk'. *Sputnik 1* was sent into orbit just a fortnight later.

This book is, above all, a miscellany. There are no themes, no theses, no coherent overviews of the course of scientific history – just a gallimaufry of oddities, from a young Charles Darwin popping a rare beetle into his mouth (with unspeakably unpleasant consequences), to the Zambian science teacher whose lofty ambition it was to send a spacegirl, a missionary and two cats to the Moon.

> Ah, but a man's reach should exceed his grasp,
> Or what's a heaven for?

One can only applaud these heroic efforts aimed at furthering the progress of science – in whatever direction, whether up, down or sideways . . .

Ian Crofton

The Ancient World

THE YEAR OF CREATION

In *The Harmony of the Four Evangelists, among themselves, and with the Old Testament* (1644), Dr John Lighfoot – who ten years later was appointed vice-chancellor of the University of Cambridge – wrote:

> And now, he that desireth to know the year of the world, which is now passing over us – this year, 1644 – will find it to be 5572 years just finished since the creation; and the year 5573 of the world's age, now newly begun, this September, at equinox.

Thus the world, according to Lightfoot, was created in 3929 BC. A few years later, Lightfoot's calculations were adjusted by James Ussher, Archbishop of Armagh, who in his *Annals of the World* (1658) worked out that the Earth had been created on 22 October, 4004 BC, at six o'clock in the evening. Before the study of geological strata began in the later 18th century, these calculations were based on the best available evidence: the generations enumerated in the Old Testament. Scientists of the calibre of Kepler and Newton made similar calculations, the former coming up with a date of 3993 BC, and the latter with 3998 BC. It is now known that the Earth is over 4.5 billion years old.

TOXIC BRONZE

The first bronze alloys were made in the Fertile Crescent of the Middle East around this time. Initially, copper was combined not with tin, but with arsenic, and the extreme toxicity of this element may account for the fact that in a number of mythologies the smith-gods – for example the Greek Hephaestus (Roman Vulcan) – are lame, muscular atrophy being one of the consequences of arsenic poisoning. It has also been suggested that the association in ancient times of smiths with lameness may be due to a custom of deliberately crippling the village smith, so that he could not leave and take his invaluable skills elsewhere. In Germanic and Scandinavian mythology, this was the fate of Weland or Wayland or Völundr the Smith, who was hamstrung on the orders of King Nidud and imprisoned on an island, where he was forced to forge armour and weapons for the king.

Apollo in Vulcan's Forge, *by Edward Francis Burney (1760–1848).*
Could the lameness of the smith-god, here seen leaning on a stick,
have been due to arsenic poisoning?

IRISH BUILD WORLD'S FIRST OBSERVATORY

The spectacular Neolithic passage grave at Newgrange in County Meath, Ireland, was constructed. In Irish mythology, it was the home of Aengus Óg, the Celtic god of love, who lived here with his lover, the swan maiden Cáer. In 1967 Professor M.J. O'Kelly observed for the first time in thousands of years that Newgrange is in fact a very accurate astronomical and calendrical instrument, designed in such a way that sunlight only penetrates to the inner chamber at dawn on the winter solstice. Today, the light first enters the chamber about 4 minutes after sunrise, and lingers for 17 minutes, but calculations that allow for the precession of the Earth indicate that 5000 years ago the light would have first appeared exactly as the Sun rose above the horizon. The famed astronomical alignments of the Great Pyramids of Egypt date from some 600 years later, while construction of the Great Temple at Karnak, aligned like Newgrange on the rising of the Sun at midwinter, did not begin until c.1375 BC.

WHY SIXTY MINUTES AND SIXTY SECONDS?

Around this time the Sumerians were using a sexagesimal (base-60) number system, which accounts for the fact that the 360 degrees in a circle are each divided into 60 minutes, as is the hour, while the minute itself is divided into 60 seconds. However, in Coordinated Universal Time (UTC), which is based on atomic clocks, leap seconds are added now and again to reflect the slowing of the rotation of the Earth, so as to keep UTC close to mean solar time. The consequence of this is that the very occasional minute (at most one a year) has either 59 or 61 seconds.

A WORM IN THE EMPRESS'S TEA

c.2600 BC
Textile technology

The discovery of silk is attributed to the legendary Chinese empress Lei Zu, who supposedly lived in the 27th century BC (although it is possible that silk was in use in China as early as 6000 BC). According to the story, a cocoon of the silkworm moth fell into the empress's tea, and unravelled in the hot water. Lei Zu found she could unwind the fine thread, which was so long it covered her entire garden. Silk is still produced by immersing the pupae in boiling water, or by sticking a needle into their bodies and unwinding the cocoon. Silk rapidly became highly valued, its attractive lustre being attributable to the triangular cross-section of the fibre, which makes it act like a prism, refracting light at different angles and so producing different colours. From the earliest times, silk was traded over great distances: silk has been found bound into the hair of an Egyptian mummy dating from the end of the 2nd millennium BC, and the Romans valued the material – which they thought came from trees – very highly.

A BRIEF HISTORY OF PI

c.1900 BC
Mathematics

The earliest written approximations of pi (the ratio of the circumference of a circle to its diameter, an irrational number that begins 3.14159 . . .) were made around this date: the Babylonians came up with $^{25}/_{8}$ (3.125), while the Egyptians opted for $^{256}/_{81}$ (3.16049 . . .). A millennium or so later, the Indian prose text *Shatapatha Brahmana* suggested $^{339}/_{108}$ (3.13888 . . .). In the 3rd century BC Archimedes used a complex geometrical method to demonstrate that pi was more than $3 + {}^{10}/_{71}$ and less than $3 + {}^{1}/_{7}$, giving an average of 3.14185. Around AD 265 the Chinese mathematician Liu Hui developed an iterative algorithm to calculate pi, and came up with an approximate value of 3.14159. Over the centuries further developments were made by Indian, Chinese and Persian mathematicians. In the early modern period, the German mathematician Ludolph von Ceulen (1540–1610) was so proud of his life's work, which largely consisted of calculating pi to 35 decimal places, that he had the number inscribed on his grave. The advent of calculus provided a new tool, and Isaac Newton calculated pi to 15 places, but thought it a waste of time, writing, 'I am ashamed to tell you to how many figures I carried these computations, having no other business at the time.' The advent of the modern computer led to pi being calculated to increasingly large numbers of decimal places. One billion was surpassed in 1989, and at the time of writing the record was over 2.5 trillion (2,576,980,370,000), achieved over a period of 73 hours 36 minutes in August 2009 on a supercomputer at the University of Tsukuba in Japan. This supercomputer, the T2K-Tsukuba, can achieve speeds of 95.4 trillion floating point operations per second, and among the 2.5 trillion digits were some surprising sequences, including 0123456789, 9876543210, 8888888888888, and even a repeat of the first 13 digits of pi: 3141592653589. *See also* 1844, 1888, 2005.

A USE FOR ACACIA AND DATES

The earliest known recipes for birth control were written down in Egypt. They describe pessaries made from cotton soaked in a paste of dates and acacia bark; the acidic qualities of these substances may have had some spermicidal effect. Carob and honey were sometimes added. It was not until much later that the condom came into use. The first description comes in a treatise on syphilis written by the Italian physician Gabriele Falloppio (1523–62), in which he describes condoms made from linen and soaked in a chemical solution as a prophylactic against the disease. They were secured over the glans by means of a ribbon. From the 17th century, condoms made from animal intestines or bladders or the finest leather were in use, although only by the well-to-do. For the first rubber condom, *see* 1855.

TIME KEEPING WITH WATER

The simplest water clocks, consisting of a stone vessel out of which water dropped at a constant rate through a hole in the bottom, were in use in Egypt and Babylon. Water clocks were also used in India and China in ancient times. The Greeks called the water clock the *clepsydra*, meaning 'water thief', and there are records from the 4th century BC of clepsydras being used in Athenian brothels to time the visits of clients. In the following century the Alexandrian physician Herophilus was using a portable clepsydra to measure his patients' pulses, while in the 1st century BC the Roman architect Vitruvius described a water clock adapted to ring an alarm. The Greeks, Romans, Byzantines, Arabs and Chinese went on to develop sophisticated gearing mechanisms and escapements, and water clocks were sometimes attached to automata, such as moving manikins or astrological models of the universe. Water clocks remained the commonest and most accurate method of keeping time until the advent of more accurate pendulum clocks in the 18th century.

POINTING THE WAY

A haematite artefact from the Olmec culture of Mesoamerica has been interpreted as an early form of compass, which would thus predate the earliest unambiguous Chinese references to a direction finder using a magnetized needle by some 2000 years. Other researchers have suggested that the Olmec find is merely a decorative object.

8th cent. BC
Public health

ADVICE REGARDING EXCRETION

In his didactic poem *Works and Days*, the Greek poet Hesiod gives the following advice:

> Do not stand upright facing the Sun when you make water, but remember to do this when he has set towards his rising. And do not make water as you go, whether on the road or off the road, and do not uncover yourself: the nights belong to the blessed gods. A scrupulous man who has a wise heart sits down or goes to the wall of an enclosed court.

c.675 BC
Cosmology

A CYLINDRICAL WORLD

The Greek pre-Socratic philosopher Anaximander (c.610–546 BC), regarded as the first scientist, produced one of the earliest maps of the world, showing Europe, Asia and Libya (Africa) grouped around the Mediterranean Sea, and surrounded by ocean. He also correctly proposed that the Earth hangs in space. He went on to suggest that the shape of the Earth was that of a cylinder, with a diameter three times its height. The inhabited world was on the top, surrounded by a circular ocean. The rest of the universe supposedly comprised a series of hollow concentric wheels filled with fire, and celestial bodies such as the Sun and the stars were what one could see of this fire through holes in the rim of the wheels.

Regarding the origins of humans, the Roman writer Censorinus recorded in the 3rd century AD that Anaximander

> ...considered that from warmed-up water and earth emerged either fish or entirely fishlike animals. Inside these animals, men took form and embryos were held prisoners until puberty; only then, after these animals burst open, could men and women come out, now able to feed themselves.

Thus Anaximander proposed some form of evolutionary process to account for the existence of human beings.

EARLY NOSE JOBS

The Indian physician Sushruta described a procedure to reconstruct noses that had been mutilated as a form of punishment. Such operations may have been carried out in India as early as 2000 BC. Sushruta's technique, using tissue bridged from the cheek or a flap of skin from the forehead, was later taken up by the Arabs, and thence made its way to Europe.

THE PERILS OF THE IRRATIONAL

Hippasus, a disciple of Pythagoras, was credited with proving that √2 is an irrational number (a real number that cannot be expressed as the ratio of two integers, such as pi). According to later legends, this so incensed his fellow Pythagoreans – who asserted that all numbers could be expressed as the ratio of two integers – that they took him out to sea and drowned him. In some accounts, it is Pythagoras himself who orders the drowning.

THE FORCES OF LOVE AND STRIFE

Death of the Greek pre-Socratic philosopher, Empedocles. He enunciated an early form of the law of the conservation of matter (first articulated in scientific terms in the 18th century), arguing that no substance is either created or destroyed, but simply changes its nature according to the ratios of four basic elements, earth, fire, water and air. These are influenced, he said, by either of two forces: the force of strife, which pulls them apart, or the force of love, which allows them to mingle. Legend says that Empedocles regarded himself as a god and threw himself into the lava-spouting crater of Mount Etna in Sicily so that people would believe, in the absence of a body, that he had ascended straight to heaven. However, his ruse was spoiled when the volcano ejected one of his bronze sandals. In the 2nd century AD the satirist Lucian of Samosata suggested that the volcano had propelled Empedocles to the Moon, where he continues to live, feeding on dew. In 2006 a dormant underwater volcano discovered off southwest Sicily was named Empedocles in his honour.

POISON GAS DEPLOYED IN PELOPONNESIAN WAR

During the siege of Athens, the Spartans set light to a mixture of charcoal, sulphur and pitch beneath the walls, with the aim of incapacitating the defenders prior to launching an assault. In the following century, Chinese texts describe defenders pumping the smoke from burning mustard into tunnels being dug by a besieging army.

HONEY MAKES MEN MAD

On his epic retreat from the disastrous Persian campaign, the Greek mercenary leader Xenophon recorded in the *Anabasis* what happened when some of his men ate the local honey:

The effect upon the soldiers who tasted the combs was, that they all went for the nonce quite off their heads, and suffered from vomiting and diarrhoea, with a total inability to stand steady on their legs. A small dose produced a condition not unlike violent drunkenness, a large one an attack very like a fit of madness, and some dropped down, apparently at death's door. So they lay, hundreds of them, as if there had been a great defeat, a prey to the cruellest despondency. But the next day, none had died; and almost at the same hour of the day at which they had eaten they recovered their senses, and on the third or fourth day got on their legs again like convalescents after a severe course of medical treatment.

The honey had been made from *Rhododendron ponticum*, the mauve-flowered species that has proved invasive in many parts of Britain. The flowers

A medieval illumination showing the toxic effects on Xenophon's men of consuming hallucinogenic honey, made from a particular species of rhododendron.

contain a chemical called acetylandromedol, which has a deleterious effect on the breathing, the nervous system and the heart.

AN OPTICAL TELEGRAPH SYSTEM

The Greek military theorist Aeneas Tacticus described a form of optical telegraph, by which simple coded messages – such as 'Cavalry arrived in the country' or 'Light-armed infantry' or 'Ships' – could be sent over long distances in a comparatively short time. Sender and receiver, positioned on separate hills, would each have an identical container of water, in each of which there floated a vertical rod, with the various agreed codes marked at different heights. The sender would raise a torch to signal the start of transmission; both sender and receiver would then open a spigot (of the same diameter) to let out water, and when the sender saw that the intended code had sunk down to the rim of the vessel, he would lower his torch. The receiver at this point would immediately turn off his spigot and read off the code on his own rod. The historian Polybius relates that such a system was used by the Carthaginians to send signals during the First Punic War (264–241 BC). In 1664 Robert Hooke presented a paper to the Royal Society in London outlining his idea for optical telegraphy, by which messages could be sent quickly over long distances via a series of towers, on top of which visual signals transmitted codes for letters or words. Hooke's ideas were not put into practice until the Revolutionary Wars of the 1790s, when the French built a network of 556 stations across their country – a network that continued in use until the 1850s. A message sent from Paris to Lille (a distance of 230 km/143 miles) would typically have taken about 32 minutes. The French initially used black and white panels, but went on to develop mechanically-controlled wooden arms, somewhat like semaphoring with flags. Other systems used panels with a number of holes that could be closed with shutters. In the UK there were lines from the Admiralty in London to key naval ports such as Portsmouth and Plymouth.

THE CALCULUS OF HAPPINESS

In *The Republic*, Plato stated a just king lives precisely 729 times more happily than a tyrant. The number 729 was of great significance to the Pythagoreans, being $3^3 \times 3^3$.

AN EARLY LIE DETECTOR

Death of the Greek physician Hippocrates of Kos, widely regarded as the Father of Medicine. He is said to have felt the patient's pulse while taking a case history to see whether he was being told the truth. Modern polygraphs also look out for an increase in pulse rate, and additionally measure other indicators such as blood pressure, respiration rate and skin conductivity.

ANIMALS BORN FROM PUTREFYING EARTH

In Book V of his *History of Animals*, Aristotle outlined his understanding of spontaneous generation (a theory first developed by various pre-Socratic philosophers):

> Some animals spring from parent animals according to their kind, whilst others grow spontaneously and not from kindred stock; and of these

instances of spontaneous generation some come from putrefying earth or vegetable matter, as is the case with a number of insects, while others are spontaneously generated in the inside of animals out of the secretions of their several organs.

Aristotle believed non-living material contained something called *pneuma* or 'vital heat', and it was this in combination with various proportions of the five elements (fire, earth, air, water and ether) that accounted for the generation of living things from inanimate matter.

ON THE ORIGINS OF EELS

Among his other observations regarding reproduction, Aristotle, having failed to detect reproductive orifices in eels, concluded that they emerged from earthworms (*History of Animals*, Book VI). In the 1st century AD, Pliny the Elder, also failing to observe reproductive passages in eels, suggested that they rubbed off particles of themselves against rocks, and that these particles grew into new eels. In the 17th century Izaak Walton, in *The Compleat Angler*, suggested that eels are generated 'as rats and mice, and many other living creatures, are bred in Egypt, by the Sun's heat when it shines upon the overflowing of the river'.

SURPRISING USES FOR ROOT VEGETABLES

In *The Masterpiece*, Aristotle lists parsnips, artichokes and turnips among other substances thought to cause erections, although he omits the root vegetable most associated with aphrodisiac qualities in the ancient world: the beetroot. Murals in a brothel in the Roman city of Pompeii (destroyed by the eruption of Vesuvius in AD 79) show people drinking what was originally thought to be red wine, but is now thought to be beetroot juice. The reputation of beetroot as an aphrodisiac is possibly due to the fact that it is high in boron, which may influence the production of sex hormones.

A LAND WITH THE CONSISTENCY OF A JELLYFISH

Some time between 330 and 300 BC the Greek explorer Pytheas, who discovered how to determine latitude by observing the altitude of the Sun, made his voyage to northwest Europe, in the course of which he circumnavigated Great Britain. Only fragments of his original account survive, in the works of later writers such as Strabo, Pliny and Diodorus. Six days sail north of Britain, Pytheas reached 'Thule', where at midsummer there was no darkness, and where 'there was no longer any proper land nor sea nor air, but a sort of mixture of all three of the consistency of a jellyfish in which one can neither walk nor sail'. This description is thought to refer to pancake ice, which forms at the edge of drift ice, and in which sea, slush and ice are mixed together.

END OF THE FLAT EARTH THEORY?

Although Pythagoras in the 6th century BC had declared, on aesthetic grounds, that all celestial objects were spherical, it was Aristotle who around this date found empirical evidence for a spherical Earth, pointing out that the further south one travels, the higher southern constellations rise above the horizon. He also noted that the shadow of the Earth on the Moon during lunar eclipses was always circular, no matter how high the Moon; only a sphere casts a circular shadow in every direction. A few decades later, Eratosthenes calculated the circumference of the Earth with some accuracy (*see* c.276 BC). By the 1st century AD, Pliny the Elder concluded that all agreed that the Earth was a sphere. Most theologians of the early Church concurred, although St Augustine (354–430) thought it ridiculous to think that the Antipodes could be inhabited:

> It is too absurd to say, that some men might have taken ship and traversed the whole wide ocean, and crossed from this side of the world to the other, and that thus even the inhabitants of that distant region are descended from that one first man [i.e. Adam].

The Christian convert Lactantius (245–325), tutor to the son of the Emperor Constantine, rejected all things pagan, including the spherical Earth theory of the Greek natural philosophers:

> Is there anyone anywhere foolish enough to think that there are Antipodeans – men who stand with their feet opposite to ours, men with their legs in the air and their heads hanging down? Can there be a place on Earth where things are upside down, where the trees grow downwards, and the rain, hail and snow fall upward? The mad idea that the Earth is round is the cause of this imbecile legend.

In 547 Cosmas Indicopleustes, a monk of Alexandria who had sailed to India, produced his *Topographia Christiana* ('Christian topography'), in which a flat Earth is modelled on the shape of the tabernacle described by God to Moses, and the heavens are arranged in the shape of a box with a curved lid.

However, these thinkers were very much in a minority, and virtually all medieval scholars who considered the matter adhered to the spherical theory, as did the scholars of the Islamic world (although in China until the 17th century the Earth was held to be flat and square). The common belief that up to the time of Columbus the vast majority of educated people in Europe believed in a flat Earth is thus fallacious, and largely attributable to Washington Irving's *The Life and Voyages of Christopher Columbus* (1828).

A revival of flat Earth theory occurred in the 19th century, when the English inventor Samuel Rowbotham (1816–84) published *Earth Not a Globe*, based on his interpretations of a number of Biblical passages. Rowbotham's teachings inspired another Englishman, Samuel Shelton, to found the Flat Earth Society in 1956. In 1972 the presidency passed to an American, Charles Kenneth Johnson, who held that the Moon landings and all the pictures of the Earth taken by orbiting astronauts were faked to lure people away from the teachings of the Bible. After Johnson's death in 2001, the society went into something of a decline, but re-launched itself in 2009.

A CURE FOR EPILEPSY

'The sound of a flute,' wrote the Greek scientist and philosopher Theophrastus, 'will cure epilepsy, and also sciatica'.

THE ORIGINAL EUREKA MOMENT

Birth of the Greek mathematician, physicist and inventor Archimedes. The best-known story concerning Archimedes relates how he solved the problem of determining the volume of an object of irregular shape. He had been asked by Hieron II, king of Syracuse, to test whether a new crown, in the shape of a laurel wreath, was indeed solid gold, as the goldsmith insisted, or whether it had been adulterated with baser metals. Archimedes was stumped, until he took a bath and noticed that the level of the water rose as he immersed himself. He realized that he could use this method to measure the volume of any object immersed in water. So, if he immersed the crown, he could work out its density by dividing its weight by its volume, and if it was not pure gold, it would have a lower density. Having come to this conclusion, Archimedes (according to the story) leapt from the bath and ran naked through the streets of the city crying '*Eureka!*' ('I have found it').

CALCULATING THE SIZE OF THE EARTH

Birth of the Greek mathematician, geographer and astronomer Eratosthenes. He famously calculated the circumference of the Earth by measuring the angle of elevation of the Sun at noon on the summer solstice, in his native Alexandria. The Sun was 7°12′ ($\frac{1}{50}$ of a full circle) south of the zenith. He knew that at the same time the Sun in the southern Egyptian city of Aswan, on the Tropic of Cancer, would be directly overhead. He thus deduced that the distance between the two cities would be $\frac{1}{50}$ of the circumference of the Earth. Estimating the distance at 5000 stadia, he therefore calculated the circumference of the Earth at just over 250,000 stadia – usually converted to 46,500 km (28,900 miles). This is some 16 per cent more than the modern measurement of 40,075 km (24,903 miles) at the Equator, but is still astonishingly accurate. Eratosthenes' logic was flawless, but the ability to measure linear distances accurately overland at that time was limited, the journey between Alexandria and Aswan being made by boat along the sinuous Nile.

CIRCULATION OF THE BLOOD – ALMOST

Death of the Greek anatomist and physician Erasistratus, famed as the inventor of the S-shaped catheter named after him. He came close to discovering the circulation of the blood when he realized that the heart was a pump rather than the seat of sensation, and distinguished between veins and arteries. He also upheld the idea that fluids when drunk passed down the oesophagus to the stomach, as against the rival theory then current, that the fluids went down the trachea (which he named) to the lungs.

HOW MANY GRAINS OF SAND TO FILL THE UNIVERSE?

In his treatise *The Sand Reckoner*, Archimedes calculated that the number of grains of sand that would be needed to fill the entire Universe was 8×10^{63} (although he did not use this modern notation). To reach this conclusion, he assumed that the Universe was spherical and centred on the Sun, and that the ratio of the diameter of the Universe to the diameter of the orbit of the Earth around the Sun equalled the ratio of the diameter of the orbit of the Earth around the Sun to the diameter of the Earth. Astronomers today realize that this is just the tiniest fraction of the size of the Universe, which has been estimated to be some 1000 billion light years across, and still expanding – rendering any attempt to calculate the number of grains of sand it might contain as fruitless an enterprise as the Walrus and the Carpenter speculating whether a beach might be cleared of its sand by seven maids with seven mops sweeping for half a year.

ARCHIMEDES' CLAW AND DEATH RAY

During the siege of Syracuse, Archimedes devised two weapons for use against the Roman ships attacking the city. The first was a giant 'claw' or

*An 18th-century engraving showing Archimedes' claw in action against
a Roman ship during the siege of Syracuse.*

'ship shaker' comprising a crane-like arm with a large grappling hook on the
end. This was dropped onto the enemy ship, and then the arm was raised,
lifting the ship into the air. Modern reconstructions of this device have
shown that it could have worked. The second weapon involved focusing
the rays of the Sun onto an enemy ship, perhaps using an array of polished
bronze or copper shields, so that the ship burst into flames. In 1747 in Paris,
the Comte de Buffon, attempting to recreate this weapon, managed to
ignite a plank of wood at a distance of 50 m (164 ft) by focusing an array
of concave mirrors on it. More recent attempts to repeat this feat using
materials that would have been available to Archimedes have had mixed
results. A 1973 experiment at a naval base near Athens succeeded in setting
fire to a plywood reconstruction of a Roman galley at a distance of 50 m
(164 ft), perhaps helped by the tar with which the wood was painted. In a
similar experiment conducted by students from the Massachusetts Institute

of Technology in 2005, a small part of a mock-up wooden ship was ignited, but only after it remained still for 10 minutes. The question arises as to why Archimedes should have bothered, given the effectiveness of the more conventional method of firing flaming missiles from a catapult.

BIG NUMBERS

In India, Jain texts were using some very large numbers. For example, 1 *koti* = 100,000,000, while 100,000,000 *koti* = 1 *pakoti*. The biggest number mentioned was *asankhyeya*, equivalent to 10^{140}, which is bigger than a googol (*see* 1938), while the number of particles in the Universe may only be 10^{80}.

ON THE AVOIDANCE OF BOOKWORMS

In his treatise *De Architectura*, Vitruvius advised building libraries so that they faced away from southerly or westerly winds, which he believed generated bookworms.

NECESSITY NOT ALWAYS THE MOTHER OF INVENTION

By this time the Maya people of Mesoamerica had invented zero, and were using a place-value system for numbers – something that the Egyptians, the Greeks and the Romans, for example, had not managed to do, making their arithmetical processes cumbersome and their mathematics limited. Several more centuries passed before zero came into use in India, from where it gradually spread to Europe via the Arabs. And yet, neither the Maya nor any other civilization of pre-Columbian America – despite their magnificent public buildings and far-flung empires – ever came up with the wheel.

THE FALSE DAWN OF THE STEAM AGE

The Greek mathematician Hero or Heron of Alexandria invented many ingenious devices, including coin-operated vending machines, a 'water organ' operated by hydraulics, and automata such as singing birds and manikins who could perform a short drama. Most notable, however, was his *aeolipile*, the first known steam engine, consisting of a boiler which fed steam up through a tubular axial shaft into a hollow sphere, on opposite sides of which were two nozzles pointing in opposite directions. As the steam was forced out of these nozzles, the sphere rotated on the axial shaft. Heron's invention was regarded as no more than an entertaining diversion, and his technology – converting steam into rotary motion – was not developed further for another 17 centuries.

Hero was also responsible for recording, in his *Stereometrica*, the first known imaginary number: the square route of -63. However, such notions were dismissed as absurdities for almost two millennia.

A CURE FOR HEAD LICE

Pliny the Elder wrote his *Naturalis Historia*, in which he recommended that the best cure for head lice was to bathe in viper's broth.

PLINY'S PEOPLE

Pliny also reported on a variety of curious subspecies of human around the world, including:

- The Arimaspians, who dwell near the source of the northeast wind, and who are distinguished by the possession of a single eye in the middle of their foreheads, and who do battle with griffons over the produce of the local goldmines.
- Another group of northerners, the cannibalistic Anthropophagi, who, to quote Philemon Holland's 1601 translation, 'have their feet growing backward, and turned behind the calves of their legs, howbeit they run most swiftly'.
- The Physilians of Africa, whose breath is 'a deadly bane and poison' to all serpents. They used this characteristic to test the fidelity of their wives: 'For so soon as they were delivered of children, their manner was to expose and present the silly [innocent] babes new born, unto the most fell and cruel serpents they could find: for if they were not right but gotten in adultery, the said serpents would not avoid and fly from them.'
- The Androgyni, also of Africa, who are 'of a double nature, and resembling both sexes, male and female, who have carnal knowledge one of another interchangeably by turns . . . Aristotle saith moreover, that on the right side of their breast they have a little teat or nipple like a man, but on the left side they have a full pap or dug like a woman.'
- The Gymnosophists of India, philosophers 'who from Sun rising to the setting thereof are able to endure all the day long, looking full against the Sun, without winking or once moving their eyes: and from morning to night can abide to stand sometimes upon one leg, and sometimes upon the other in the sand, as scalding hot as it is.'

- Also in India are to be found 'a kind of men with heads like dogs, clad all over with the skins of wild beasts, who in lieu of speech use to bark: armed they are and well appointed with sharp and trenchant nails: they live upon the prey which they get by chasing wild beasts, and fowling.'
- Another native people of the subcontinent were the Monoscelli, 'that have but one leg apiece, but they are most nimble, and hop wondrous swiftly'. During the hottest weather they would lie on their backs and 'defend themselves with their feet against the Sun's heat'.

The death of Pliny the Elder while investigating the eruption of Vesuvius, as depicted by Delacroix in the Palais-Bourbon, Paris.

- 'Again, beyond these Westward,' writes Pliny, 'some there be without heads standing upon their necks, who carry eyes in their shoulders.'
- Then, around the source of the Ganges, are the Astomes, who have no mouths: 'They live only by the air, and smelling to sweet odours, which they draw in at their nostrils.'
- At the foot of the Himalaya there was a race of pygmies, much troubled, according to Homer, by attacks from cranes, against whom they would mount armed expeditions, born along on the backs of rams and goats. Having reached the cranes' breeding ground at the seaside, 'they make foul work among the eggs and young cranelings newly hatched, which they destroy without all pity'.

A VICTIM OF SCIENTIFIC CURIOSITY

When the eruption of Vesuvius began, Pliny the Elder was staying on the other side of the Bay of Naples, but when he saw what was happening, he ordered one of the ships of which he had charge to take him across the bay to Pompeii. Despite the constant rain of hot ash and pumice, they continued, until they found themselves trapped on a lee shore. Pliny disembarked, but in a while, his nephew Pliny the Younger reported, he was overcome by thick fumes, and could not rise from the ground, forcing his companions to abandon him. Two days later his body was found, without a mark on it. It may be that the fumes exacerbated an existing heart condition, but if he was still alive when last seen, he would shortly afterwards have become just one of many victims of the second phase of the eruption, the pyroclastic flow – a cloud of gas and rock fragments heated to 1000°C and travelling at speeds of 700 kph (450 mph). Anybody breathing this gas would be immediately incinerated from the inside.

ON THE DETUMESCENT EFFECTS OF PIGEONS

In his *Epigrams*, the Roman poet Martial declared 'Wood pigeons check and blunt the manly powers; let him not eat the bird who wishes to be amorous.'

THE ORIGINAL BOTTLE BLONDES

Roman hairdressers (*ornatrices*) used pigeon excreta mixed with ashes to bleach the hair of their clients.

TOADS AND DRAGONS DETECT EARTHQUAKES

Zhang Heng, the Chinese scientist, inventor and poet, demonstrated his greatest invention, the first seismometer, at the imperial Han court. It consisted of a bronze urn inside which was a complex mechanism involving a swinging pendulum, levers and cranks. If a distant tremor or earthquake should occur, a ball would be dropped from one of eight dragon's mouths into the mouth of one of eight toads, each representing a different point of the compass. Having established that a disaster had occurred, the authorities could then send aid in the right direction.

Zhang Heng was also credited with building the first water-powered armillary sphere (a moving model of the heavenly bodies, conceived as a celestial sphere) and an odometer, which marked how far a carriage had travelled by causing a mechanical figure to beat a drum (for one *li* covered) or a gong (for ten *li*).

A reproduction of Zhang Heng's pioneering seismometer, presented to the Chinese emperor in AD 132.

AD 138
Reproductive science

ON THE USE OF SNEEZES

In his *Gynæcology*, the Greek physician Soranus of Ephesus gave the following advice on contraception:

> The woman ought, in the moment during coitus when the man ejaculates his sperm, to hold her breath, draw her body back a little so that the semen cannot penetrate the *os uteri*, then immediately get up and sit down with bent knees and, in this position, provoke sneezes.

WOUNDS AS WINDOWS INTO THE BODY

At the age of 28, the Greek physician Galen became physician to the gladiators of the High Priest of Asia in Pergamon. Dissecting human cadavers was against Roman law, so Galen drew often erroneous inferences about human anatomy by looking at the insides of dead – and living – pigs and monkeys. His job with the gladiators, however. gave him some insights into human anatomy, and he referred to the wounds he treated as 'windows into the body'. Galen's teachings on medicine, anatomy and physiology were to dominate thinking on these matters in Europe and the Islamic world until the beginnings of the Scientific Revolution (*see* 1543).

C. AD 290
Mathematics

A GRAVE RIDDLE

Death of the Alexandrian mathematician Diophantus. A Greek collection of puzzles from the 5th century gives him the following riddling epitaph:

> This tomb holds Diophantus. Ah, how great a marvel! The tomb tells scientifically the measure of his life. God granted him to be a boy for one-sixth of his life, and adding a twelfth part to this, he clothed his cheek with down. He lit him the light of wedlock after a seventh part, and five years after his marriage he gave him a son. Alas, late-born wretched child! After attaining the measure of half his father's life, chill Fate took him. After consoling his grief by the study of numbers for four years, Diophantus ended his life.

From this we can work out that, where x is the age of Diophantus at his death

$$x = \tfrac{1}{6}x + \tfrac{1}{12}x + \tfrac{1}{7}x + 5 + \tfrac{1}{2}x + 4$$

So it turns out that Diophantus died at the age of 84 – although this cannot be relied on from a historical point of view.

A CURIOSITY REGARDING THE NUMBER SIX

Death of the Neoplatonist philosopher Iamblichus. He had proposed that if you chose any three consecutive numbers the largest of which is divisible by 3, add them together, then add the digits of the answer, and then add the digits of *that* answer, and so on, you will eventually end up with the answer 6.

MORE ON THE NUMBER SIX

Death of St Augustine of Hippo. He had been impressed with the number 6, the first perfect number, i.e. it is equal to the sum of its factors $(1 + 2 + 3)$. He wrote:

> Six is a number perfect in itself . . . God created all things in six days because this number is perfect. And it would remain perfect, even if the work of six days did not exist.

The number 6 is also equal to the product of its factors $(1 \times 2 \times 3)$, and no other number is both the sum and the product of the same three numbers.

EFFIGIES TYCHONIS BRAHE O. F.
ÆDIFICII ET INSTRUMENTORVM
ASTRONOMICORVM STRVCTORIS
Aᵒ DOMINI 1587 ÆTATIS SVÆ 40

The Middle Ages and the Renaissance

On the environmental effects of menstrual flow ∗ Byzantine flamethrowers ∗ The origin of gibberish? ∗ Some early birdmen ∗ The vegetable lamb of tartary ∗ Comet a source of tears ∗ An eternally turning wheel? ∗ Journey to the centre of the Earth ∗ Pus not laudable after all ∗ A use for wolf's penis ∗ An early blood transfusion ∗ On the dangers of figs ∗ The gruesome route to the truth ∗ Copernicus the fool ∗ The mystery of the swallow solved ∗ Carried to the Moon by geese

ON THE ENVIRONMENTAL EFFECTS OF MENSTRUAL FLOW

In his popular encyclopedic compendium *Etymologiae*, St Isidore of Seville described the deleterious effects of menstrual blood:

> On contact with this gore, crops do not germinate, wine goes sour, grasses die, trees lose their fruit, iron is corrupted by rust, copper is blackened. Should dogs eat any of it, they go mad. Even bituminous glue, which is dissolved neither by iron nor by water, polluted by this gore, falls apart by itself.

Such misogynistic superstitions masquerading as science were not new. Pliny the Elder, in his *Natural History* (c.AD 77), claimed that 'On the approach of a woman in this state, milk will become sour, seeds which are touched by her become sterile, grafts wither away, garden plants are parched up, and the fruit will fall from the tree beneath which she sits.'

THE END OF ANCIENT LEARNING?

Legend has it that the conquering Arab army burnt down the great Library of Alexandria, repository of much of the learning (scientific and otherwise) of the ancient world. According to the story, when the Arab commander Amr ibn al'Aas asked the Caliph Umar what to do with the library, the latter ordered him to burn all the books, 'For they will either contradict the Koran, and are thus heretical, or will agree with it, in which case they are superfluous.' It is said that there were sufficient numbers of books to heat bath water for the entire army for six months. However, the library had suffered several fires in the past, and many scholars do not credit the story of the 642 destruction.

BYZANTINE FLAMETHROWERS

According to the Byzantine chronicler Theophanes, an architect from Heliopolis in Egypt called Kallinikos invented a naval weapon that became known as 'Greek fire'. This was an inflammable mixture projected via a tube or siphon at enemy vessels using pressurized pumps. Anna Komnene,

Greek fire in action, as illustrated in the Skylitzes Codex, a Byzantine illuminated manuscript from the 12th century.

the 12th-century Byzantine princess and historian, has left the following account of the use of Greek fire by her father, the Emperor Alexios I Komnenus, against a Pisan fleet:

> As he knew that the Pisans were skilled in sea warfare and dreaded a battle with them, on the prow of each ship he had a head fixed of a lion or other land-animal, made in brass or iron with the mouth open and then gilded over, so that their mere aspect was terrifying. And the fire which was to be directed against the enemy through tubes he made to pass through the mouths of the beasts, so that it seemed as if the lions and the other similar monsters were vomiting the fire.

Greek fire was credited with saving Constantinople from two Arab sieges, and such was its importance that later Byzantine chroniclers said the method for making it was first revealed by an angel to Constantine, the first Christian emperor, in the 4th century. Its exact recipe is lost, but it may have included some or all of various ingredients, such as sulphur, quicklime (calcium oxide), naphtha (unrefined petroleum), and nitre (potassium nitrate or sodium nitrate), with resin used as a flammable thickening agent. It could burn on the surface of the sea, but could be extinguished if covered by sand (which deprived it of oxygen). Also effective in putting out the fire were old urine or strong vinegar, both of which presumably broke down its chemical composition.

c. 721
Chemistry

THE ORIGIN OF GIBBERISH?

Birth of Geber, the Latinized version of Jabir ibn Hayyan, the Arab or Persian polymath who is regarded as the 'Father of Chemistry'. Among his many achievements was the development of alchemy into an experimental science – he understood the workings of acids and alkalis (he named the latter) and discovered that boiling wine released a flammable vapour – alcohol.

His inventions included fire-proof paper, ink that could be read in the dark, and a (presumably oil-based) substance that made cloth waterproof and kept iron free from rust. He wrote his alchemical texts in a highly esoteric language, incomprehensible to non-initiates, and it has been suggested that his name gave rise to the English word 'gibberish'. (Another theory derives the word from Gibraltar, where the natives baffled outsiders by mixing Spanish and English in the same sentence.)

THE ORIGIN OF ALGEBRA

The Persian mathematician Al-Khwārizmī wrote his treatise on solving polynomial equations, *Hisab al-jabr w'al-muqabala*, the title of which gave rise to our word 'algebra'. The Arabic word *al-jabr* can mean bone-setting, or reunification, or mathematical reduction.

THE BOOK OF INGENIOUS DEVICES

The Banu Musa brothers, three Persian scholars who lived in Baghdad, published their *Book of Ingenious Devices*. This included descriptions of scores of automata and other mechanical devices, some dating back to the Greeks, and others of their own invention. Among the latter are a number of mechanical musical instruments, including a hydropowered organ that could play interchangeable cylinders automatically, and an automatic flute player, perhaps the first programmable machine.

SOME EARLY BIRDMEN

In the Emirate of Córdoba in southern Spain, the Moorish inventor Abbas Ibn Firnas, then aged 65, made an early attempt at manned flight, as described many centuries later by the historian Ahmed Mohammed al-Maqqari (d.1632):

> Among other very curious experiments which he made, one is his trying to fly. He covered himself with feathers for the purpose, attached a couple of wings to his body, and, getting on an eminence, flung himself down into the air, when according to the testimony of several trustworthy writers who witnessed the performance, he flew a considerable distance, as if he had been a bird, but, in alighting again on the place whence he had started, his back was very much hurt, for not knowing that birds when they alight come down upon their tails, he forgot to provide himself with one.

Abbas Ibn Firnas has been honoured on a number of postage stamps, and has a crater on the Moon named after him.

Other early would-be aeronauts include an English monk called Eilmer, who in 1010 launched himself from the tower of Malmesbury Abbey in Wiltshire and glided some 180 metres (200 yards) on homemade wings before landing and breaking both legs. Also worthy of note is the flight in 1507 of Father John Damian, a notorious alchemist and quack, from the walls of Stirling Castle in Scotland. Damian, whose effort was witnessed by King James IV, had intended flying all the way to France, but instead he landed in a dunghill and broke his thigh bone. It has been pointed out that his wings must have been reasonably effective, for Stirling Castle stands on a sheer rock and if he had not glided some distance, perhaps 800 metres (half a mile), he would have been killed. Damian himself blamed his failure on his choice of materials: he should not have used the feathers of hens, he said, as such birds 'covet the middens [dunghills] and not the skies'.

It is assumed that all these efforts sought to imitate gliding birds, for humans do not have the necessary musculature to beat artificial wings. As the biologist J.B.S. Haldane pointed out in his 1927 essay, 'On Being the Right Size':

> An angel whose muscles developed no more power weight for weight than those of an eagle or a pigeon would require a breast projecting for about four feet [1.2 m] to house the muscles engaged in working its wings, while to economize in weight, its legs would have to be reduced to mere stilts.

994

Toxicology / pathology

THINGS FALL APART

Thousands died in Aquitaine, southwest France, after consuming rye infected with a mould, *Claviceps purpurea*, which contains a range of toxic ergot alkaloids. This form of fungal poisoning, then called St Anthony's fire and now known as ergotism, causes convulsions, diarrhoea, vomiting, mania, psychosis, hallucinations and gangrene of the limbs, followed by death. In an earlier outbreak, in 857, a chronicler reported that:

> A great plague of swollen blisters consumed the people by a loathsome rot, so that their limbs were loosened and fell off before death.

The condition was called St Anthony's fire because it was thought that praying to St Anthony would effect a cure; around 1095 the Order of St Anthony was founded to care for those suffering from the disease. It has been suggested that the symptoms of bewitchment – convulsions, hallucinations, crawling sensations on the skin, and so on – exhibited by the accused in the Salem witchcraft trials of 1692–3 indicates they may have been suffering from ergotism.

THE VEGETABLE LAMB OF TARTARY

Reports filtered into Europe of a plant that grew in Central Asia, known as the *Agnus scythicus* ('Scythian lamb') or *Planta tartarica barometz* ('vegetable lamb of Tartary', *barometz* being the local word for 'lamb'). The fruit of this strange plant was a lamb, which was attached by a stem or umbilical cord. Its mobility thus constrained, the lamb fed on the grass growing about the plant, and when this was all consumed both lamb and plant died. The myth may well have been a means of explaining cotton, and the plant was also identified with *Cibotium barometz*, the woolly fern, native to parts of China and the Malay Peninsula.

BRIGHT LIGHTS

In China, the Middle East and Europe observations were made, low in the southern sky, of a supernova – the great explosion that occurs when certain kinds of star die. The 1006 supernova was said to be as much as three times the size of Venus, and with one-quarter of the brightness of the Moon. The Chinese reported that it was half the size of the Moon, and so bright that objects on the ground could be seen at night by its light. Astrologers believed it boded no good, with war and famine being sure to follow. The supernova of 1054 seen by Chinese, Japanese, Persian and Arab astronomers was perhaps even brighter, being visible in daylight for 23 days. Its remains are still visible as the Crab Nebula. In the historical record, only two further supernova have been visible with the naked eye: those that occurred in 1572 and 1604, which were recorded by Tycho Brahe and Johannes Kepler respectively. Tycho Brahe correctly concluded that the appearance and disappearance of the supernova of 1572 showed that the stars were not fixed and unchanging, as the ancient Greek astronomers had believed, but were born and ultimately died.

LOVE SICK

Death of the great Persian physician Avicenna (Ibn Sina), whose writings on medicine became standard works in the Muslim world and in Europe down to the 18th century. Among his many innovations was his understanding of the effect of emotions on physiological processes, and the story is told of how he felt the pulse of a very sick man while listing the names of various places and people. He noticed that the patient's pulse quickened when certain names were mentioned, and from this deduced that his patient was in love with a certain young woman. He immediately recommended marriage as the cure, and soon after the wedding the man had completely recovered.

COMET A SOURCE OF TEARS

(20 March) The appearance of Halley's Comet was taken as an ill omen in England. 'You've come, have you?' said Eilmer, the flying monk of Malmesbury (*see under* 875), who may have witnessed the comet's previous appearance in 989. 'You've come, you source of tears to many mothers, you evil. I hate you! It is long since I saw you; but as I see you now you are much more terrible, for I see you brandishing the downfall of my country. I hate you!' (This at least is how William of Malmesbury, writing in the following century, recorded Eilmer's words.) Seven months after the appearance of

the comet the English were decisively defeated by the Normans under William the Conqueror at the Battle of Hastings, and their king, Harold II, was killed. It was the end of Anglo-Saxon England. *See also* 1835.

AN ETERNALLY TURNING WHEEL?

The Indian mathematician and astronomer Bhaskara II came up with the first known design for a perpetual-motion machine, in the form of a wheel that had containers of mercury around the rim. This, he said, would rotate ceaselessly, as one side of the wheel was always heavier than the other. There is no record that he succeeded in building a working model. We now know that any kind of perpetual motion machine is an impossibility, as it would break one or other of the laws of thermodynamics.

QUICKLIME IN THE EYES

(24 August) At the Battle of Sandwich, off the Kent coast, an English fleet defeated a French invasion force under the command of the master pirate Eustace the Monk. A key moment came when the English threw pots of powdered quicklime (calcium oxide) onto the deck of the French flagship. The French crew, thus blinded, were unable to repel the English men-at-arms who then boarded the ship and slaughtered all whom they came across. Eustace the Monk was found cowering in the bilges, and, his offer of a 10,000 mark ransom being refused, his head was cut off and displayed on the tip of a lance.

The next development in chemical warfare came some centuries later, when Leonardo da Vinci suggested that naval vessels use mangonels to catapult pots of 'chalk, fine sulphide of arsenic and powdered verdigris' onto the enemy galleys.

SPLITTING HILLS

Death of Michael Scot, the Scottish mathematician and scholar, who worked in Spain and at the Sicilian court of the Emperor Frederick II. Scot translated into Latin several works of Aristotle on biology and astronomy from the Arabic in which they had been preserved. His own works largely concerned astrology and alchemy, and he earned a widespread reputation as a wizard – hence his appearance in the eighth circle of Hell alongside other magicians and soothsayers in Dante's *Inferno*. Among the many legends that grew up around him were that he employed a demon to split the Eildon Hills in the Scottish Borders into three separate peaks, and that in Cumbria he literally petrified a coven of witches, the result being the stone circle known as Long Meg and Her Daughters.

JOURNEY TO THE CENTRE OF THE EARTH

Death of the French Dominican friar Vincent of Beauvais, author of the *Speculum Maius* ('great mirror'), the main encyclopedia used in the Middle Ages. The *Speculum* is largely a summary of the science, natural history, geography and history known in western Europe at this date. Vincent also included some of his own speculations, such as the method of communication employed by angels (a form of intelligible speech), and what would happen to a stone dropped down a hole bored right through the Earth (it would stop in the middle, he decided).

DREAMS OF FLIGHT

Roger Bacon, the English Franciscan friar and polymath, sent his *Opus Majus* to the pope. In this work, which ranges over science, mathematics, grammar and philosophy, Bacon anticipated later inventions such as microscopes, telescopes, spectacles, steam ships and flying machines. Many legends subsequently grew up around Bacon, such as that he constructed a brazen head that could not only talk, but also answer any question put to it. This makes an appearance in Robert Greene's play *Friar Bacon and Friar Bungay* (c.1589), in which it only manages to make three utterances – 'Time is,' 'Time was,' and 'Time is past' – before falling to the floor and shattering into pieces.

PUS NOT LAUDABLE AFTER ALL

The Lombard surgeon William of Saliceto boldly contradicted the teachings of Galen (*see* AD 157), long regarded as sacrosanct, by claiming that the presence of pus in a wound was not in fact a healthy sign. Galen had concluded that the formation of 'laudable pus' ('*pus bonum et laudabile*') in a wound was part of the body's healing process, and applied a variety of substances, including dung, to encourage this formation. However, he did observe that spreading infection in a wound could often result in systematic sepsis and death. Nearly a century after William of Saliceto, Galen's view was still being upheld by the great surgeon Guy de Chauliac in his *Chirurgia magna* (1363). However, in due course there arose among surgeons a new adage, to wit: '*Ubi pus, ibi evacua*' (Latin, 'Where there is pus, there evacuate it').

A USE FOR WOLF'S PENIS

(15 November) Death of Albertus Magnus, the German scientist and theologian known as 'Aristotle's Ape', such was his devotion to the ancient Greek. He also came up with his own theories – for example, in *De Animalibus* he asserted:

> If a wolf's penis is roasted in an oven, cut into small pieces, and a small portion of this is chewed, the consumer will experience an immediate yen for sexual intercourse.

He also noted that starfish was such a strong aphrodisiac that the subject was likely to end up ejaculating blood. In such an eventuality, the subject should eat something cooling, like lettuce. Albertus Magnus was not short of helpful medical advice. For example, if one were to find oneself infested with lice, the quickest cure was to smear one's skin with the excrement of an elephant.

FROM TWO TO FOUR EYES

Tomaso da Modena painted a portrait of Cardinal Hugh de Provence, the first known visual depiction of the use of eyeglasses (spectacles) used for reading. The making of the first such glasses, around 1284, has been credited to Salvino D'Amato of Florence; although this has been disputed, it is likely that the first spectacles were indeed made in Italy between 1280 and 1300. The first spectacles for short-sightedness may have been made by Nicholas of Cusa in the 15th century, while the first bifocals were invented – and worn – by Benjamin Franklin in 1784.

A portrait of Cardinal Hugh de Provence painted in 1352 by Tomaso da Modena. This is the first known depiction of eyeglasses.

FORTY DAYS OF ISOLATION

The port of Marseille instituted a rule that all ships coming from plague-infected areas must be kept in isolation for 40 days. The word 'quarantine' comes from Italian *quaranta*, '40', although the first quarantine recorded in Europe, imposed in Ragusa in 1377, prescribed an isolation period of only 30 days. Six centuries later, on their return from the Moon, the astronauts from the Apollo programme (up to and including *Apollo 14*) were quarantined for a period to avoid the introduction of extraterrestrial microbes into the Earth's biosphere.

TESTING FOR VITAL SIGNS

John of Mirfield (1362–1407), librarian and physician at St Bartholomew's Hospital in London, offered the following advice:

> If there is any doubt as to whether a person is or is not dead, apply lightly roasted onions to his nostrils, and if he is alive, he will immediately scratch his nose.

AN EARLY BLOOD TRANSFUSION

On his deathbed, if we are to believe a hostile chronicler called Stefano Infessura, Pope Innocent VIII was given the blood of three boys in an attempt to revive him. According to Infessura, the blood was administered to the pontiff by mouth. All four died in the process.

THE TUMULTUOUSLY TALENTED FUNDAMENT OF
THEOPHRASTUS PHILIPPUS AUREOLUS BOMBASTUS VON HOHENHEIM

Birth in Switzerland of Phillip von Hohenheim, the renowned physician, botanist, alchemist, astrologer and occultist. He later adopted the name Theophrastus Philippus Aureolus Bombastus von Hohenheim and awarded himself the title Paracelsus, the name by which he is generally known. It means 'equal to (or greater than) Celsus', alluding to the Roman author of *De Medicina* and other encyclopedic works. Paracelsus was not afflicted by false modesty, and famously remarked:

> All the universities and all the old writers put together are less talented than my arsehole.

He challenged the traditional view, based on Hippocrates and Galen, that disease was caused by imbalances in the body's four 'humours', and believed instead that it was caused by outside agents – an anticipation of Pasteur's germ theory, but for Paracelsus the responsible agents were poisons originating in the stars, rather than microorganisms. A pioneer of modern chemistry, he has been seen as the father of toxicology and pharmacology, with his observation that 'All things are poison and nothing is without poison, only the dose permits something not to be poisonous.' He held the chair of medicine at the University of Basel, but after less than a year was obliged to leave, having scandalized his colleagues by publicly burning the books of Galen and Avicenna, claiming that his shoe buckles were more learned than either. He then set out on extensive travels through Europe, North Africa and the Middle East, in search of old manuscripts containing esoteric knowledge, before being appointed physician to Duke Ernst of Bavaria shortly before he died in mysterious circumstances at the White Horse Inn in Salzburg in 1541. Some say he died of natural causes, others that he was stabbed in a brawl.

A woodcut, attributed to Paracelsus, illustrating in the three heads at the top the alchemical principles of salt, sulphur and mercury.

WHAT'S WRONG WITH LISA?

The enigmatic expression on the face of the *Mona Lisa* – Leonardo da Vinci's celebrated portrait of Lisa Gherardini, wife of Francesco del Giocondo – has long baffled those who have gazed on that famous face. But in 2010 Vito Franco, a medical scientist at the University of Palermo, offered a different interpretation of her appearance. What he noticed was a xanthelasma – a yellowish and well-defined subcutaneous accumulation of cholesterol – in the hollow of her left eye. Although not harmful or painful in itself, a xanthelasma may indicate high levels of cholesterol in the blood. Franco also noticed a lipoma – a benign tumour composed of fatty tissue – on one of Lisa's hands. Examining other masterpieces from the Renaissance, Franco has pointed out that the models for Botticelli's *Portrait of a Youth* and Parmigianino's *Madonna with the Long Neck* may both have suffered from Marfan's syndrome, a genetic disorder that affects the connective tissues. Sufferers are typically tall, and have unusually long limbs and long, thin fingers. As for the depiction of Michelangelo as the philosopher Heraclitus in the centre foreground of Raphael's *The School of Athens*, Franco believes that his swollen knees may indicate an excess of uric acid, which can cause painful kidney stones – and this in turn might explain his gloomy expression and despondent demeanour, resting his head on his hand. The condition may have resulted from the long months he worked on the Sistine Chapel, living off nothing but bread and wine.

A MELANCHOLY OBSESSION

The German artist Albrecht Dürer published his famous engraving *Melencolia I*, which includes the magic square on the page opposite. Each row and column adds up to 34, as do both diagonals, each of the four quadrants, the central four squares, the four corner squares of the 4 × 4 grid, as well as the corner squares of each of the contained 3 × 3 grids. The numbers adjacent to each corner clockwise (3 + 8 + 14 + 9) also add up to 34, as do the numbers adjacent to each corner anticlockwise

16	3	2	13
5	10	11	8
9	6	7	12
4	15	14	1

(5 + 15 + 12 + 2). Further patterns, all adding up to 34, can also be extracted. The two numbers in the centre of the bottom row give the year of publication, 1514, while the numbers either side, if denoting the order of the letters of the alphabet, give A and D, the initials of the artist.

A NEW ARK

(20 February) In 1499 the German mathematician and astronomer Johannes Stöffler, a professor at the University of Tübingen, predicted that on this date the world would be overwhelmed by a great deluge. He based his prediction on the fact that 20 planetary conjunctions would take place in 1524, 16 of them in a 'watery sign', i.e. Pisces. As the date approached, boat-builders across Europe benefited from full order books, and a German nobleman, Count von Iggelheim, launched a three-storey ark on the Rhine. When on 20 February a few spits and spots of rain began to fall, there was a rush for Iggelheim's ark, and in the ensuing riot the count was stoned to death. Stöffler's reputation was damaged when 1524 turned out to be a year of drought, and no one took his recalculated date of 1528 too seriously. He died of the plague in 1531 – not, as he had predicted, by being hit by a falling body.

ON THE DANGERS OF FIGS

In his *Dietary of Health*, Andrew Boorde warned against the consumption of figs, as they 'doth provoke a man to sweat, wherefore they doth engender lice'.

THE GRUESOME ROUTE TO THE TRUTH

The Flemish anatomist Andreas Vesalius published *De humani corporis fabrica*, in which he overthrew many of the erroneous teachings of the Greek physician Galen regarding human anatomy (*see* AD 157). Galen had largely relied on animal dissection, and his teachings had been regarded as sacrosanct throughout the Middle Ages. But Vesalius was determined to dissect human rather than animal corpses, and early in his career, in order to obtain a complete human skeleton, he had stolen the putrefied cadaver of a criminal hanging from a gibbet outside the city of Louvain. On another occasion, while a wrongdoer was being disembowelled while still alive, he managed to get hold of 'the still-pulsating heart with the lung and the rest of the viscera'. In 1539, after he had moved to Padua, a judge who had become interested in his work made available to him the corpses of newly executed criminals, and even arranged matters so that executions would take place just before Vesalius's anatomy lessons. Vesalius's work aroused considerable hostility. The pious were shocked to learn that men had the same number of ribs as women – for did not the Bible tell us that Eve was created out of Adam's rib? For their part, the traditionalists were so upset when Vesalius enumerated some 200 anatomical errors in Galen's teaching that they insisted that the human body must have changed since the days of the ancient Greek.

COPERNICUS THE FOOL

Publication of *De revolutionibus orbium coelestium*, in which Copernicus proposed that the Earth moves round the Sun, and not vice versa, so contradicting Ptolemy and Christian dogma. 'This fool wishes to reverse the entire science of astronomy,' scoffed Martin Luther, 'but sacred scripture tells us that Joshua commanded the Sun to stand still, and not the Earth.' For its part, the Roman Catholic Church only removed *De revolutionibus* from its *Index* of prohibited books in 1835.

THE MYSTERY OF THE SWALLOW SOLVED

Olaus Magnus, Archbishop of Uppsala in Sweden, reported that in the seas to the north, fishermen would on occasion haul up their nets to find fish and swallows 'hanging together in a conglomerated mass'. Thomas Pennant, in his *British Zoology* (1766), says that the archbishop:

> . . . very gravely informs us, that these birds are often found in clustered masses at the bottom of the Northern Lakes, mouth to mouth, wing to wing, foot to foot; and that they creep from the reeds in autumn, to their sub-aqueous retreats . . . That the good Archbishop did not want credulity, in other instances, appears from this, that after having stocked the bottoms of the lakes with birds, he stores the clouds with mice, which sometimes fall in plentiful showers on Norway and neighbouring countries.

Others said that swallows hibernated, or changed species, or flew to the Moon, the return journey taking some 60 days.

INNUMERABLE SUNS, INNUMERABLE EARTHS

The Italian monk and Neoplatonist philosopher Giordano Bruno published *De l'infinito universo e mondi* ('on the infinite universe and worlds'), in which he went further than Copernicus in claiming that there was not just one sun at the centre of the universe, but that the universe was full of them: 'Innumerable suns exist; innumerable earths revolve around these suns in a manner similar to the way the seven planets revolve around our sun. Living beings inhabit these worlds.' For this and other heresies Bruno was burnt at the stake in 1600.

A TREATMENT FOR SORE EYES

The Portuguese-born Rodrigo Lopez was appointed physician to Queen Elizabeth I. Among his recommendations was that sore eyes might be soothed by bathing them in urine. In 1593 he was accused (almost certainly falsely) of plotting to poison the queen, and on 7 June 1594 he was hanged, drawn and quartered.

AMOROUS APPLES?

In his *Haven of Health*, Thomas Cogan stated that 'Apples are thought to quench the flame of Venus.' However, when during the Commonwealth period of the following century the Puritans banned people from dancing round the phallic maypole, they resorted to dancing around apple trees instead.

CARRIED TO THE MOON BY GEESE

Francis Goodwin, Bishop of Llandaff in Wales, began to write *The Man in the Moon: or, A Discourse of a Voyage Thither, by Domingo Gonsales*, which was not published until 1638, after Goodwin's death. Goodwin was clearly aware of Copernicus's heliocentric universe, and by and large his narrative proceeds in obedience to the laws of physics, as then understood, with a goodly admixture of fantasy. Gonsales is carried to the Moon in a flying machine powered by 40 wild geese, which migrate to the Moon each winter, and notices that as he travels further from the Earth he becomes lighter, and then becomes heavier again as he approaches the Moon. This is inhabited by a variety of intelligent, human-like creatures, who live in a kind of utopia. They ensure that this idyllic state of affairs continues by identifying any moral or physical defects at birth, and any infant suffering from either is promptly exiled from the Moon and sent to North America.

THE BOY WITH THE GOLDEN TOOTH

Professor Jakob Horst of the Academia Julia in Helmstedt, Lower Saxony, examined the mouth of a seven-year-old boy, Christoph Müller, about whom extraordinary reports had been circulating, to the effect that he had a golden tooth. When Horst looked into the boy's mouth, he indeed saw the tooth, and rubbing it with a touchstone confirmed that it was indeed made of gold. Horst ascribed the golden tooth to the astrological circumstances of the boy's birth, on 22 December 1585, when the planetary alignments, he concluded, would have accentuated the heat of the Sun, causing part of the boy's jaw bone to turn to gold. Sadly for Horst's thesis, the march of time – in the form of continued mastication and frequent testing by touchstone – wore down the gold, showing that it had just been a thin skin of metal closely fitted over the normal tooth underneath. This is the first known example of a moulded gold crown. When the deception was exposed, the skilful artificer who had fitted the crown disappeared; as for the unfortunate boy, he was thrown into prison.

The 17th Century

The moons of Jupiter do not exist * Space travel and witchcraft * A fatal frozen chicken * A use for human brain * Why the Earth does not move * Weighing one's own stools * An immaculate conception? * Spontaneous generation of mice * The seat of the soul? * Sixteen horses defeated by vacuum * A cure for impotence * Wasting time on weighing air * Schoolboy flogged for not smoking * Sheep's blood transfused into man * On the origins of maggots * A new element extracted from urine * The ceiiinosssttuu puzzle * A complaint against scientists * Showers of frogs

FLIGHTS OF FANTASY

The English physician William Gilbert published *De Magnete*, in which he concluded from his experiments that the Earth is magnetic, and has an iron core, so accounting for the operation of the compass. Gilbert's studies led him to condemn all suggestions that a perpetual-motion machine could be driven by magnets: 'May the gods damn all such sham, pilfered, distorted works,' he declared, 'which do but muddle the minds of students.'

It was Gilbert's *De Magnete* that inspired the floating island of Laputa in Jonathan Swift's *Gulliver's Travels* (1726): Laputa is a disc, largely composed of metal, four and half miles (7.2 km) in diameter, and with a six-yard (5.5 m) bipolar magnet buried inside it. It floats over the Earth's surface at a place where there is a particularly intense magnetic field, but, by turning the magnet, it can be steered to different locations, both horizontally and vertically. The Laputans view everything through the lens of science and mathematics, and have very little interest in human affairs down below. The one Laputan who does show an interest is regarded as 'the most ignorant and stupid person among them'.

THE MOONS OF JUPITER DO NOT EXIST

(January) Using one of his new telescopes, Galileo discovered Jupiter's four largest satellites, Io, Europa, Callisto and Ganymede. His fellow astronomer Francisco Sizzi (who was later to discover the annual variations in the Sun's sunspot activity) was dismissive. 'Jupiter's moons are invisible to the naked eye,' he argued, 'and therefore can have no influence on the Earth, and therefore would be useless, and therefore do not exist.'

In Jonathan Swift's novel Gulliver's Travels *(1726), the island of Laputa floats above the Earth by means of magnetic forces.*

SPACE TRAVEL AND WITCHCRAFT

Around this time the astronomer Johannes Kepler, renowned for his laws of planetary motion, circulated a manuscript entitled *Somnium* ('the dream'). In this early work of science fiction, Kepler explores the new heliocentric universe of Copernicus from the perspective of the Moon, to which his hero, Duracotus, travels from the Earth. Kepler anticipates Newton's concept of gravity, describing how the journey of Duracotus becomes easier because the body 'escapes the magnetic force of the Earth and enters that of the Moon, so that the latter gets the upper hand'. He goes on to describe the freezing cold nights and the fiercely hot days experienced on the Moon, and the lunar creatures that are adapted to these stark conditions. Hanging in the sky above them, waxing and waning, is the Earth, appearing many times the size of our terrestrial view of the Moon. Although *Somnium* is a meticulous attempt to understand the new 'celestial physics', it became a key piece of evidence in the trial of Kepler's mother for witchcraft. In 1615, Katharina Kepler was accused by a woman in a financial dispute with Kepler's brother Cristoph of having given her a potion that made her ill. A charge of witchcraft followed, and the prosecution produced a garbled rewrite of *Somnium*, in which the mother of the narrator learns the secrets of space travel from a demon. Kepler was assiduous in his mother's defence, and she was eventually acquitted in 1621. The original text of *Somnium*, accompanied by 223 explanatory footnotes provided by the author following his mother's release from prison, was published in 1634, four years after Kepler's death. Shortly prior to this latter event, Kepler had written his own epitaph:

> I used to measure the heavens,
> Now I measure the shadows of the Earth.
> Although my mind was heaven-bound,
> The shadow of my body lies here.

ON THE PREVENTION OF INEBRIATION

Death of the lawyer and horticulturalist Sir Hugh Platt. As well as offering much good advice on gardening, he suggested the following method to avoid getting excessively drunk:

> Drink first a good large draught of salad oil, for that will float upon the wine which you shall drink, and suppress the spirits from ascending into the brain. Also, what quantity soever of new milk you drink first, you may well drink thrice as much wine after, without danger of being drunk.

Sir Hugh realized there could be a down side, adding 'But how sick you shall be with this prevention, I will not here determine.'

ON THE POTENTIAL OF STEAM

Publication of *Raison des forces mouvantes* (later translated into English as *Of Forceful Motions: Description of divers machines both useful and joyful*) by the French hydraulic engineer and garden designer Salomon de Caus. In this he described a fountain in which water is forced upward by the expansion of steam, and this has led some to credit de Caus with inventing the steam engine. It is possible that three years earlier de Caus had built such a fountain at Richmond to entertain the young heir to the thrones of England and Scotland, Henry, Prince of Wales, who died of typhoid in November 1612. Although he did not show how steam power could be harnessed to provide rotary motion, de Caus seems to have seen the potential, which some years later he enthusiastically outlined to Cardinal Richelieu. 'To listen to him,' snorted the cardinal, 'you would fancy that with steam you could navigate ships, move carriages. In fact, there's no end to the miracles which, he insists upon it, could be performed.' So absurd did Richelieu find de Caus's 'fancies' that he had him confined to the lunatic asylum at Bicêtre, where he died in 1626. It has been suggested that de Caus's fate was not in fact prompted by his 'fancies', but by a beautiful mistress angered by his jealousy and possessiveness. According to the story, she wrote to Richelieu – whom she numbered amongst her former lovers – asking him to find a way to rid her of 'this embarrassing lunatic'.

AN ARITHMETICAL CURIOSITY

Galileo showed that one-third bears an interesting relation to the sequence of odd numbers:

$$1/_3 = (1 + 3) / (5 + 7) = (1 + 3 + 5) / (7 + 9 + 11) =$$
$$(1 + 3 + 5 + 7) / (9 + 11 + 13 + 15) \ldots$$

THE CONTINENTAL JIGSAW

Francis Bacon, credited as the begetter of the modern scientific method, noted that the coastlines on either side of the Atlantic, particularly the eastern protuberance of Brazil and the West African shore along the Gulf of Guinea, formed a neat fit. It was an observation that lay undeveloped until 1912, when a German meteorologist called Alfred Wegener, having noted the same resemblance and having assembled a body of additional circumstantial evidence, floated the theory of continental drift. This met with derision from most geologists. But the theory of plate tectonics, which emerged in the 1960s, provided a mechanism by which continental drift occurs – not to mention explaining earthquakes, volcanoes, sea-floor spreading, the creation and destruction of rocks, and much else. There was some residual opposition, for example from the International Stop Continental Drift Society, whose founder, geologist John Holden, explained 'This organization valiantly campaigned to stop continental drift and sea-floor spreading. Its efforts to reunite Gondwanaland, have, however not yet achieved the desired results.'

CONCERNING SWORDS AND KNIVES

In his *Medicus Microcosmus*, Daniel Beckher, professor of medicine at the University of Königsberg, advised that in the case of a large knife or sword wound, the weapon that had inflicted the wound should be 'anointed' every day, and that it should be 'kept in pure linen and in a warm place but not too hot, nor squalid, lest the patient should suffer harm'.

Despite this apparent belief in sympathetic magic, Beckher deserves a place in the history of surgery for overseeing one of the first attested instances of a successful gastrotomy (surgical incision into the stomach). The patient was a farmer, Andreas Grünheide, who one morning in the summer of 1635 woke up feeling sick, so to make himself vomit he poked the handle of a long knife down his throat. Unfortunately, he let go and the knife became stuck. To try to dislodge it he stood on his head, but to no avail. He then quaffed some beer to lubricate his throat. Even more unfortunately, the knife then slipped all the way down his oesophagus to his stomach. Beckher brought his patient before the full medical faculty and the consensus was that surgery should be attempted. The operation was carried out on 9 July by Daniel Schwabe, and the knife successfully removed. Restored to health, Grünheide returned to his farm, and six years later got married.

A FATAL FROZEN CHICKEN

During an unseasonably cold and snowy spring, Francis Bacon (*see* 1620) was riding by coach up Highgate Hill, to the north of London. According to John Aubrey in his frequently unreliable *Brief Lives*, compiled later in the century, Bacon stopped to conduct an experiment: having purchased a chicken, he killed and gutted it and then stuffed it with snow with his own hands to see whether the cold would retard putrefaction. As a result, he caught a chill, and was obliged to take shelter in the Earl of Arundel's house nearby. From his bed, Bacon wrote his last letter to the absent earl:

> I was likely to have had the fortune of Caius Plinius the Elder, who lost his life by trying an experiment about the burning of the mountain Vesuvius [*see* AD 79]. For I was also desirous to try an experiment or two, touching the conservation and induration of bodies. As for the experiment itself, it succeeded excellently well; but in the journey (between London and Highgate) I was taken with such a fit of casting, as I knew not whether it were the stone, or some surfeit, or cold, or indeed a touch of them all three. But when I came to your Lordship's house; I was not able to go back, and therefore was forced to take up my lodging here, where your housekeeper is very careful and diligent about me.

Although the housekeeper supplied Bacon with a warming pan, Aubrey tells us the bed was damp, and Bacon succumbed to 'suffocation' (i.e. pneumonia) and died on 9 April. In their 1998 biography of Bacon, however, Lisa Jardine and Alan Stewart propose a different cause of death. Bacon may have been experimenting on himself, they suggest, inhaling nitre or opiates in an attempt to alleviate his ill health – but with fatal effects.

A USE FOR THE HUMAN BRAIN

1631
Medicine

Death of Johannes Hartmann, professor of 'chymiatrie' at the University of Marburg. He was the first professor of chemistry anywhere, and his interests were directed towards *materia medica* (pharmacology). For example, in his *Praxis Chymiatrica* he recommends 'spirit of human brain' as a cure for epilepsy. An alternative cure involved administering a powder prepared from the livers of live green frogs – although, to be effective, the preparation could only be carried out in the months of May, June or July.

WHY THE EARTH DOES NOT MOVE

1633
Astronomy

Scipio Chiaramonti, professor of mathematics and philosophy at the University of Pisa, put forward what he regarded as an irrefutable argument: 'Animals, which move, have limbs and muscles. The Earth has neither limbs nor muscles, therefore it does not move.'

ON THE IDENTIFICATION OF WITCHES

1634
Anatomy

Four of the accused in the notorious Pendle Witches case were brought to London for a physical examination by a jury of surgeons and midwives under the direction of William Harvey. Harvey was a private physician to King Charles I and celebrated in the annals of science as the first to correctly describe the circulation of the blood. It was believed by many at the time that witches could be identified by the presence of certain marks, such as a teat in an untoward place, which, it was asserted, they would use to suckle the Devil. Harvey and his team applied a more rigorous scientific approach to the procedure, reporting as follows:

> On the bodies of Janet Hargreaves, Frances Dickinson and Mary Spencer nothing unnatural neither in their secrets or any other parts of their bodies, nor anything like a teat or mark nor any sign that any such thing hath ever been.
>
> On the body of Margaret Johnson we find two things may be called teats, the one between her secrets and the fundament on the edge thereof, the other on the middle of her left buttock. The first in shape like to the teat of a bitch, but in our judgements nothing but the skin of the fundament drawn out as it will be after the piles or application of leeches. The second is like the nipple or teat of a woman's breast but of the same colour with the rest of the skin without any hollowness or issue for any blood or juice to come from thence.

The four accused were pardoned by the king, and the boy who had been the chief prosecution witness admitted that his stories of covens and the metamorphosis of humans into animals and back had been entirely fabricated.

WEIGHING ONE'S OWN STOOLS

(22 February) Death of the Italian physician Sanctorius of Padua, who conducted one of the first quantitative studies of human metabolism. Over a period of three decades he meticulously weighed everything he consumed, both food and drink, and everything he excreted, both urine and faeces. Finding that the output weighed considerably less than the input, he attributed the difference to 'insensible perspiration'.

AN IMMACULATE CONCEPTION?

In Grenoble, France, Magdeleine d'Auvermont, wife of Jérôme de Montléon, Seigneur d'Aiguemère, was brought before the court, having giving birth to a son called Emmanuel. Jérôme's relatives wanted the child to be declared illegitimate, on the grounds that Jérôme, a captain of horse, had been away at the wars in Germany for four years. In defence of the legitimacy of her child and her own honour, Magdeleine swore that she had known no other man, but that in a dream her husband had made love to her, and the next morning she knew she was with child. The boy was born nine months later. Expert witnesses in the shape of four midwifes testified that they too had borne children without having had sex, and the possibility of this was vouchsafed by four physicians from the University of Montpellier, adherents of the doctrine of parthenogenesis or spontaneous generation (*see* 350 BC). On 13 February the court ruled that Emmanuel was indeed the legitimate heir to the Seigneur d'Aiguemère and all his estates.

FERMAT ON FORM

The French mathematician Martin Mersenne (*see under* 1903) wrote to his fellow mathematician Pierre de Fermat asking what the ratio of $2^{36} \times 3^8 \times 5^5 \times 11 \times 13^2 \times 19 \times 31^2 \times 43 \times 61 \times 83 \times 223 \times 331 \times 379 \times 601 \times 757 \times 1201 \times 7019 \times 823{,}543 \times 616{,}318{,}177 \times 100{,}895{,}598{,}169$ was to the sum of its proper divisors. Fermat replied that the answer was 1 to 6, and for good measure pointed out that the prime factors of the last number, 100,895,598,169, were two prime numbers, namely 112,303 and 898,423. Quite how Fermat worked this out remains a mystery – comparable to his claim to have demonstrated the truth of his own last theorem (which was not proved until 1994).

SPONTANEOUS GENERATION OF MICE

(30 December) Death of the Flemish chemist, physiologist and physician, Jan Baptist van Helmont. Among his many interesting experiments was the one in which he stuffed a dirty shirt into a hole in a barrel that had been filled with grains of wheat. After about 21 days, he reported, there was a noticeable change in the smell, and the products of decomposition had apparently penetrated the husks of the wheat and transformed the grains into mice.

In another experiment, van Helmont planted a willow sapling in an earth-filled pot, and for five years added nothing to the pot but water. He then dug up the tree and weighed it. At 77 kg (169 lb 3 oz) it was now 30 times as heavy as when he had planted it, while the soil weighed virtually the same. He concluded that the great increase in weight of the wood, bark and roots had been 'produced by water alone'. His reasoning seemed infallible, but, being unaware of photosynthesis, he did not realize the contribution made by atmospheric carbon dioxide, not to mention oxygen, light and trace elements from the soil.

THE SEAT OF THE SOUL?

In his *Passions of the Soul*, the French philosopher René Descartes described the pineal gland as 'the seat of the soul', believing it to be the place where the mind interacted with the body. In 2007, a research team from the National Taiwan University in Taipei reported that brain scans indicated that the area around the pineal gland becomes active when people meditate. The co-leader of the team suggested that their findings 'might have profound implications in the physiological understanding of mind, spirit and soul'.

Other scientists expressed scepticism that the pineal gland has any other function than the secretion of melatonin, a hormone that controls our biological clock.

SIXTEEN HORSES DEFEATED BY VACUUM

To demonstrate the efficacy of his new vacuum pump, Otto von Guericke joined two hollow copper hemispheres together and evacuated the air from the interior. He then attached each hemisphere to a team of eight horses, and then got the two teams to pull in opposite directions. The horses were unable to separate the hemispheres.

A CURE FOR IMPOTENCE

(10 January) Death of the English herbalist and physician Nicholas Culpeper. For any man unable to provide his wife with 'due benevolence' in the marital bed, Culpeper suggested he pass water through his wife's wedding ring. A similar cure was popular in France, with the alternative of urinating through the keyhole of the church in which they had been married. In his painting *Venus and Cupid*, dating from the 1520s, Lorenzo Lotto depicts Cupid urinating through a bridal wreath held up by Venus, the stream landing on her lap. The whole painting is intended as a blessing on a wedding, and includes other symbols of fecundity and fidelity.

Culpeper had some alternative suggestions for pepping up the jaded appetite. In his *Complete Herbal* (1653) he stated: 'Asparagus . . . being taken fasting several mornings together, stirreth up bodily lust in man or woman.'

INVENTION OF THE RAMJET?

Cyrano de Bergerac published *Les États et empires de la lune* ('the states and empires of the Moon'). In this early work of science fiction he describes how the hero, Dyrcona, is taken to the Moon by a machine powered by fireworks. In the course of his detailed description the narrator provides what Arthur C. Clarke regarded as an anticipation of the modern ramjet, a

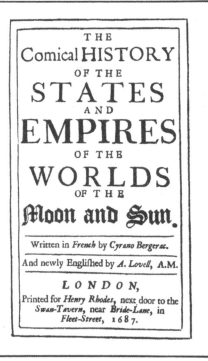

The engraved frontispiece and title page from the first English edition of Cyrano de Bergerac's pioneering work of science fiction.

form of jet engine that uses the forward motion of the aircraft to compress incoming air:

> I foresaw very well, that the vacuity . . . would, to fill up the space, attract a great abundance of air, whereby my box would be carried up; and that proportionable as I mounted, the rushing wind that should force it through the hole, could not rise to the roof, but that furiously penetrating the machine, it must needs force it upon high.

The first patent for a ramjet was granted to the French inventor René Lorin in 1913, but he failed to build a prototype. The first flight of an aircraft powered by a ramjet took place in the Soviet Union in 1939.

WASTING TIME ON WEIGHING AIR

(1 February) In his diary, Samuel Pepys noted that King Charles II 'mightily laughed' at the scientists for spending their time 'only in weighing air, and doing nothing else'. He was referring to Robert Boyle's experiments in measuring the mass, volume and pressure of gases, which resulted in Boyle's law.

SCHOOLBOY FLOGGED FOR NOT SMOKING

During the Great Plague, a pupil at Eton was 'never whipped so much in his life as he was one morning for not smoaking'. At that time smoking tobacco was thought to be an effective way of preventing infection – along with carrying a posy of roses, or sitting next to a stinking privy, or consuming woodlice, viper's fat or crab's eyes.

1667
Haematology

SHEEP'S BLOOD
TRANSFUSED INTO MAN

(23 November) At a meeting of the Royal Society, Samuel Pepys witnessed an experiment conducted by Richard Lower, author of *Tractatus de corde* ('treatise on the heart'), whereby about 'eight or nine ounces' of the blood of a sheep was transfused into a 'poor and debauched man . . . cracked a little in the head', with the hope that the procedure might 'have a good effect upon him as a frantic man by cooling his blood'. Remarkably, the subject,

A 1671 Dutch engraving showing blood being transfused from a dog into a man. Such procedures normally proved fatal to both parties.

one Arthur Coga, survived. 'The patient found himself very well upon it,' reported Henry Oldenburg, secretary of the Royal Society, 'his pulse better than before, and so his appetite.' Similar experiments subsequently carried out in France proved fatal. Coga, a highly eccentric scholar who had fortified himself with 'a cup or two of sack' before the procedure, believed he would come to no harm, the blood of the lamb being symbolic of the blood of Christ.

THE BEGINNING OF THE TWO CULTURES?

Thomas Sprat published his *History of the Royal Society*, in which he outlined the sort of language appropriate to science, which should dispense with 'luxury and redundance of speech' and 'reject all amplifications, digressions and swellings of style'. Instead, scientists should adopt 'a close, naked, natural way of speaking; positive expressions, clear senses; a native easiness: bringing all things as near the mathematical plainness, preferring the language of artisans, countrymen and merchants, before that of wits or scholars'.

ON THE ORIGINS OF MAGGOTS

Francesco Redi published his *Experiments on the Generation of Insects*, in which he showed, by comparing pieces of meat placed in open and gauze-covered containers, that maggots did not, as had previously been believed, emerge spontaneously from rotting meat. He went on to observe the metamorphosis of maggots into flies, and noted that if living flies were placed in sealed jars with dead animals, maggots appeared.

A NEW ELEMENT
EXTRACTED FROM URINE

1669
Chemistry

The German merchant and amateur alchemist Hennig Brand spent years combining various materials in pursuit of the 'philosopher's stone', the substance that would, it was believed, turn base metals into gold. In one experiment he filled 50 buckets with human urine and left them until they grew putrid and 'bred worms'. He then heated the liquid to boiling, and when it was reduced to a paste, he heated the residues until the retort glowed red hot. Eventually strange glowing vapours emerged, which later condensed to a liquid, and then to a white solid, which glowed so brightly that Brand could read by its light at night. He called this substance 'phosphorus', from the Greek word *phosphoros*, meaning 'light-bringing'.

THE CEIIINOSSSTTUU
PUZZLE

1676
Physics

In a postscript to *A Description of Helioscopes and some other Instruments*, the English physicist Robert Hooke stated that he had discovered 'the true theory of elasticity or springiness', but, jealous that rivals might steal his ideas, he encrypted the principle in an anagram: 'ceiiinosssttuu'. Two years later, once he was more confident of his results, he published the promised explanation in *Lectures de potentia restitutiva, or of spring*, in which he revealed the solution to the anagram: '*Ut tensio sic vis*':

> That is, the power of any spring is in the same proportion with the tension thereof: that is, if one power stretch or bend it one space, two will bend it two, and three will bend it three, and so forward.

In other words, the extension of a spring is directly proportional to the force applied (provided that the force does not exceed the elastic limit) – a discovery since known as Hooke's law.

FIRST ESTIMATE OF THE SPEED OF LIGHT

Working in Paris as assistant to Giovanni Domenico Cassini, the Danish astronomer Ole Rømer observed that the times between the eclipses of the moons of Jupiter grew shorter as the Earth approached Jupiter, and lengthened as the Earth moved away. Rømer concluded that the speed of light is finite, and that it takes 22 minutes to traverse the diameter of the orbit of the Earth. This gives a speed of 226,666 km (140,533 miles) per second – a figure certainly in the same region as the modern value of 299,792 km (185,871 miles) per second. The discrepancy is largely due to an inaccuracy in the then-accepted estimate of the diameter of the Earth's orbit.

A COMPLAINT AGAINST SCIENTISTS

An anonymous pamphlet was published attacking scientists as dilettantes:

> We prize ourselves in fruitless curiosities; we turn our lice and fleas into bulls and pigs by our magnifying glasses; we are searching for the world in the Moon with our telescopes; we send to weigh the air on the top of Tenerife . . . which are voted ingenuities, whilst the notions of trade are turned into ridicule or much out of fashion.

TAKING THE BISCUIT

Gottfried Leibniz published his first paper on calculus, but subsequently Isaac Newton claimed that *he* had invented the method, in 1666, and that Leibniz must have stolen his idea, having been shown an unpublished manuscript of his work. There ensued a long drawn-out and bitter quarrel between the two men and their supporters, stretching into the first two decades of the following century. The Royal Society in London set up a committee to investigate the dispute. The report it issued in 1713 came out in favour of Newton – not surprisingly, given that it was written by Newton himself. Thus it was somewhat disingenuous (if not downright hypocritical) of Newton to subsequently write to one of Leibniz's allies, the mathematician Johann Bernouilli, with the following protestations:

> Now that I am old, I have little pleasure in mathematical studies, and I have never tried to propagate my opinions over the world, but I have rather taken care not to involve myself in disputes on account of them.

For his part, in 1716, a few months before his death, Leibniz explained why he had kept silent:

> In order to respond point by point to all the work published against me, I would have to go into much minutiae that occurred thirty, forty years ago, of which I remember little: I would have to search my old letters, of which many are lost. Moreover, in most cases I did not keep a copy, and when I did, the copy is buried in a great heap of papers, which I could sort through only with time and patience. I have enjoyed little leisure, being so weighted down of late with occupations of a totally different nature.

Today, the consensus is that both men invented calculus independently, but it is Leibniz's form of notation that has proved the most useful.

As a footnote, it is pleasing to mention that the names of the two adversaries are preserved in the names of two biscuits. Leibniz-Keks are plain butter biscuits, and have been made since 1891 by the Bahlsen company of Hanover, of which city Leibniz was a prominent resident. Fig Newtons are a type of fig roll first made by the Kennedy Biscuit Company in 1891; in the 1950s they were advertised on television by a cowboy singing 'Yer darn tootin', I like Fig Newtons.' (Only a pedant of the worst sort would point out that Fig Newtons are named, not after the great scientist, but after the town of Newton in Massachusetts.)

SHOWERS OF FROGS

A shower of frogs fell on Lord Aston's bowling green at Tixall, according to Robert Plot in his *Natural History of Staffordshire*. In the annals of meteorology this is by no means a unique occurrence. To mention just a few similar instances: during a thunderstorm in 1881, several tons of hermit crabs and periwinkles fell from the sky on to the city of Worcester, while in 1948 a shower of herrings landed on a golf course in Bournemouth. More recently, in 1987, a shower of albino frogs fell on Stroud, Gloucestershire. A possible explanation of such phenomena was offered in the early 19th century by the French physicist André Marie Ampère (*see* 1836), who, noting that at certain times toads and frogs congregate in large numbers in the countryside, suggested that they might be swept up into the air by a violent wind and carried some distance before, so to speak, precipitating. Given that in most instances the creatures involved are aquatic, subsequent hypotheses have largely involved the agency of waterspouts.

The 18th Century

A whiff of brimstone * Newton
predicts end of world * The flying
boy with sparks drawn out of his
nose * Jumping monks * The witch
of Agnesi * A new terror of death *
Predicting the date of one's own death
* Electrical stimulation of turnips *
A mathematical proof of the existence
of God? * On the dangers of living
too fast * Wind to be rendered as
agreeable as perfumes * The first
glimmer of a black hole * Frenetic
frogs and convulsing corpses * A
poetical big crunch * Giant-headed
creatures on the Sun

A WHIFF OF BRIMSTONE

In his *Gazophylacii Naturae & Artis*, William Petiver illustrated a hitherto unknown butterfly, a specimen of which had been sent to him by the butterfly collector William Charlton. The specimen resembled the sulphur-yellow Common Brimstone (*Gonepteryx rhamni*), except that it had black spots and blue moons on its wings. When it was examined by Carl Linnaeus in 1763 he declared it a new species, naming it *Papilio ecclipsis*. Linnaeus also referred to a specimen in a collection of North American butterflies. Who it was who realized that *P. ecclipsis* was in fact a Common Brimstone with black spots painted on its wings is unclear, but when the hoax was drawn to his attention, Dr Edward Grey, Keeper of Natural Curiosities at the British Museum in which the Petiver specimen was held, was apparently so enraged that he 'indignantly stamped the specimen to pieces'. Some say it was the Danish entomologist Johan Christian Fabricius who exposed the fake in 1793, while others credit John Curtis, author and illustrator of the masterly *British Entomology* (1824–39). A further candidate is the lepidopterist William Jones, who some believe to be responsible for the two specimens now preserved in the Linnean Collections of the Linnean Society of London. Certainly the handwritten label on one of these includes the cryptic Latin comment '*Rhamni arte pictus!*' ('Rhamni artfully painted'), followed by the name 'Jones'.

NEWTON PREDICTS END OF WORLD

1704 Eschatology

Isaac Newton predicted in a letter that the world would end in 2060, based on what he believed were coded messages in the Bible. The end would come along the lines described in the apocalyptic Book of Revelation, with plagues, fires, the Battle of Armageddon between good and evil, 'the ruin of the wicked nations, the end of weeping and of all troubles'. Newton believed these events would be followed by the Second Coming of Christ and the thousand-year reign of the saints, among whom he numbered himself. Newton offered some comfort to his contemporaries regarding the date of the fate of the world: 'It may end later,' he wrote, 'but I see no reason for its ending sooner.'

Newton was usually more precise in matters of timing, although notoriously absent-minded, as recorded by the Irish poet Thomas Moore in the 1820s: 'He insisted that his breakfast egg must cook exactly five minutes; on one occasion the maid entered the kitchen to find Newton before the stove, thoughtfully looking at the egg, which rested in his hand, while his watch lay in the saucepan of boiling water.'

A UNIT IS BORN

Birth of the distinguished French scientist Claude Émile Jean-Baptiste Litre, the son of a wine merchant. He is best-known for proposing the unit of volume measurement that bears his name, which was incorporated into the International System of Units after his death in 1778. Little is known of his life, but he apparently had a daughter called Millie. This at least was the assertion by Kenneth Woolner of the University of Waterloo, Ontario, Canada, who in the April 1978 issue of *Chem 13 News* used this fictional character to justify the practice of using 'L', rather than 'l' (which in many fonts is indistinguishable from the numeral '1') as the symbol for 'litre'. Normally the International System of Units only allows capital letters as symbols in cases where the unit is named after an actual historical person – for example, N for newton, W for watt, J for joule.

SMALLPOX PARTIES

In a letter from Adrianople to her friend Sarah Chiswell, Lady Mary Wortley Montagu, wife of the British ambassador to the Ottoman Empire, described the method of partial immunization against smallpox then practised among the Turks:

> Every autumn in the month of September, when the great heat is abated, people send to one another to know if any of their family has a mind to have the smallpox. They make parties for this purpose, and when they are met (commonly 15 or 16 together) the old woman comes with a nutshell full of the matter of the best sort of smallpox and asks what veins you please to have opened. She immediately rips open that you offer to her with a large needle (which gives you no more pain than a common scratch) and puts into the vein as much venom as can lie upon the head of her needle, and after binds up the little wound with a hollow bit of shell, and in this manner opens four or five veins . . .

Lady Mary determined to introduce the practice back in England, if she could find any doctor there who 'had virtue enough to destroy such a considerable branch of their revenue for the good of mankind'. In 1721 she had her three-year-old daughter inoculated against smallpox by the eminent surgeon Charles Maitland. This aroused much interest, and Princess Caroline, future queen of George II, decided to have her own daughters inoculated. However, as the procedure involved the insertion of pus from a smallpox victim into a scratch on the skin, Princess Caroline insisted that half a dozen prisoners at Newgate destined for the gallows try it out first. Seeing that they came to no harm, she proceeded to have her own daughters inoculated. The prisoners escaped hanging.

The procedure was not in fact harmless, 1 in 50 of those being inoculated with smallpox dying of the disease; the others were all capable of spreading it to other people. It took Edward Jenner to develop the technique of vaccination, by which people could be made immune to smallpox by inoculating them with the far less dangerous cowpox (*see* 1788).

THE FLYING BOY WITH SPARKS DRAWN OUT OF HIS NOSE

Stephen Gray, the English dyer and amateur 'experimental philosopher', came up with the concept of electrical conductivity, attributing electrical

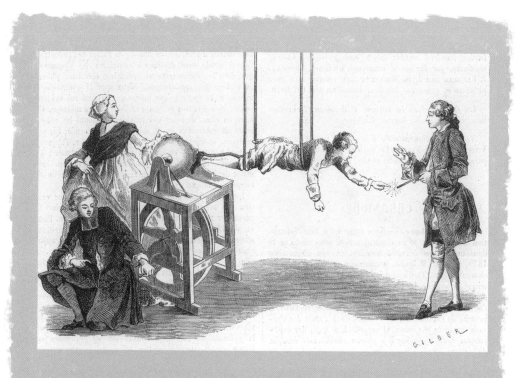

Jean-Antoine Nollet (see 1746) performs a variation on Stephen Gray's 'Flying Boy' experiment, demonstrating how electricity is conducted through the human body.

effects to the 'communication' of the electrical 'virtue'. Always of modest means, he was admitted this same year as a pensioner to the Charterhouse in London, where he went on to put on demonstrations of electrical phenomena that had as much to do with show business as science. His most notable turn was 'The Flying Boy', which involved suspending a Charterhouse schoolboy by silk cords and charging him up by rubbing him with a glass rod until his hands attracted bits of paper and chaff, and sparks could be drawn from his nose. The aim of these 'philosophical fireworks' was, in the words of Gray's assistant Anna Williams (the daughter of another Charterhouse pensioner),

> . . . to break the sleep of elemental fire;
> To rouse the pow'rs that actuate Nature's frame.

William Stukeley, one of the first biographers of Newton, called Gray 'the father, at least first propagator, of electricity'. The Royal Society made Gray a fellow in 1733, and awarded him the first Copley Medal for scientific achievement. Three years later Gray died destitute, and is thought to have been buried in a common grave, alongside other pauper pensioners of the Charterhouse.

THE DOWNSIDE OF HYDROTHERAPY

1737
Public health

The anonymous author of 'The Diseases of Bath: A Satire' described the state of the waters at Bath, then the most fashionable resort of the English *haute monde*. He starts by describing how full the baths were with the foul by-products of the diseased: filth, lepra, scabies, peeling scales . . .

Nor is this the greatest grievance in the flood;
The worst I scarcely wish were understood:
All (from the porter to the courtly nymph)
Pay liquid tributes to the swelling lymph . . .
Hence mad and poisoned from the bath I fling
With all the scales and dirt that round me cling:
Then looking back, I curse that jakes obscene,
Whence I come sullied out who entered clean.

REPLENISHING THE CRISPY FIBRES

The author of *The Ladies' Physical Directory* offered his Prolifick Elixir, a 'Powerful Confect' and 'Stimulating Balm', which, he assured his readers, would

> fortify the Nerves, increase the Animal Spirits, restore a juvenile Bloom, and evidently replenish the crispy Fibres of the whole Habit, with a generous Warmth and Moisture.

'Animal spirits' was the key phrase, as impotence could arise if they were deficient, or ceased 'to flow in such abundance to the particular Muscles, and other Parts administering to Generation'.

A REMARKABLE RECOVERY

(24 November) A youth called William Duell was hanged at Tyburn for 'occasioning the death of Sarah Griffin, at Acton, by robbing and ill-treating her'. Subsequently, his body was taken to Surgeons' Hall to be dissected by the anatomists. What happened next is taken up by the *Newgate Calendar*:

. . . but after it was stripped and laid on the board, and one of the servants was washing it, in order to be cut, he perceived life in him, and found his breath to come quicker and quicker, on which a surgeon took some ounces of blood from him; in two hours he was able to sit up in his chair, and in the evening was again committed to Newgate, and his sentence, which might be again inflicted, was changed to transportation.

JUMPING MONKS

Jean-Antoine Nollet, the Abbot of the Grand Convent of the Carthusians in Paris, and also a pioneer in the study of electricity, devised an experiment to prove his theory that electricity travels far, and so fast as to be almost instantaneous. To this end, he lined up 200 of his monks, linking each pair with a wire 3 m (25 ft) long. He then attached a Leyden jar (a device for storing electricity which he named, but did not invent) to the first monk, and noted with pleasure how each monk yelped and writhed simultaneously as the current flowed through them. He conducted similar experiments in front of King Louis XV using 180 guardsmen, and the king himself participated in another such demonstration. However, subsequent experiments found that sometimes the current fizzled out after just a few people. Some suggested that electricity could not pass through an impotent or castrated man, or a frigid woman, but when the experiment was tried on a number of castrati – male singers who had had their testicles removed prior to puberty – the subjects reacted violently to the shock. The real reason that the experiment sometimes failed was almost certainly related to the dampness of the ground on which the subjects stood – the damper the ground, the more conductive it would be, allowing the electricity to earth down the subjects' legs.

THE WITCH OF AGNESI

The Italian mathematician Maria Agnesi (1718–99) of the University of Bologna described the plane curve:

$$x^2y = 4a^2(2a - y)$$

She called it the *versiera*, from the Italian for a sheet, the rope used to control a sail (from Latin *vertere*, 'to turn'). However, when her description was translated into English, *la versiera* was confused with *l'avversiera*, meaning 'the woman contrary to God', and so the curve has come to be known as 'the witch of Agnesi'. After her father's death in 1752, Agnesi abandoned mathematics and devoted her energies to the succour of the poor of Milan.

A HITHERTO UNKNOWN DANGER OF YELLOW FEVER

Two doctors in Kingston, Jamaica, disagreed so violently about the origin of yellow fever that they fought a duel in which both were killed.

RESTORING THE HONOUR OF CERTAIN WOMEN

The Royal Society in London received a lengthy letter from a certain Abraham Johnson, entitled *Lucina Sine Concubita* (Latin, 'pregnancy without sex'). In it, Johnson declared that he had used 'a wonderful, cylindrical, catoptrical, rotundo-concavo-convex machine' to capture 'animalcula' floating in the air, which, when viewed under a microscope, had the form of miniature men and women. It was these free-floating animalcula that Johnson blamed for those not uncommon instances of pregnancy where the woman swears she has remained chaste and a virgin. The letter was a spoof, perpetrated by Sir John Hill, but intended as a serious satire of the theory of spermism, by which it was held that every individual on Earth originally existed as a homunculus inside Adam's testicles, a homunculus passed from generation to generation via the male line. Even the detection of spermatozoa by Anton van Leeuenhoek using his microscope did not scotch the theory, and in 1694 Nicolaas Hartsoeker, in his *Essai de dioptrique*, published an image of a miniature human curled up inside a sperm. The question then arose as to why Divine Providence should allow so many of these homunculi to be wasted in the course of each ejaculation. In response to this difficulty, the philosopher and mathematician Gottfried Leibniz advocated the theory of panspermism, by which the sperm were not wasted, but carried by the wind until they should find something suitable to fertilize. Hence Sir John Hill's roving, rampant animalcula, presenting a danger to maidenly chastity everywhere. As a means of proving his hypothesis, Hill/Johnson suggested to the Royal Society that an edict be issued, banning copulation for a year throughout Great Britain. Any pregnancies that occurred would be proof that panspermism was a reality.

A NEW TERROR OF DEATH

With the intention of 'better preventing the horrid crime of murder', the Murder Act 1752 decreed that 'some further terror and peculiar mark of infamy be added to the punishment', to wit, the body of no executed murderer was to be buried, but rather left hanging in chains – or subjected to a public dissection. This new provision increased the supply of corpses to the medical schools, but supply could not keep up with demand, so unscrupulous operators dubbed 'Resurrectionists' or 'body snatchers' would dig up recently buried bodies and sell them to the anatomists. One such body was that of the novelist Laurence Sterne, author of *Tristram Shandy*, who died in 1768. But when the cadaver was wheeled into the anatomy theatre at Cambridge University, the students immediately recognized Sterne from his much reproduced portrait by Sir Joshua Reynolds, and refused to proceed with the dissection.

Some were too impatient to wait for their subjects to die of natural causes: in a down-at-heel lodging house in Edinburgh in 1827–8 William Burke and William Hare notoriously hastened 17 people towards their end via inebriation followed by smothering. They then sold the corpses to Dr Robert Knox of the Edinburgh Medical School – giving rise to the rhyme:

> Up the close and doun the stair,
> But an' ben wi' Burke an' Hare.
> Burke's the butcher, Hare's the thief,
> Knox the boy that buys the beef.

The subsequent outcry led to the passage of the Anatomy Act 1832, by which physicians, surgeons and medical students were permitted to dissect any corpse unclaimed after death, such as the bodies of those who died in prison or the workhouse. In addition, if a person agreed to donate the corpse of their next of kin, their own burial expenses would be met by the institution receiving the body. *See also* 1846.

GIVE US BACK OUR ELEVEN DAYS!

Britain and its empire adopted the Gregorian calendar. This had been introduced by Pope Gregory XIII in the 16th century, but rejected as a Romish innovation by several Protestant countries in Europe. Most of these gradually adopted the Gregorian calendar, but it was not until 1752 that Britain – now 11 days behind most of the rest of Europe – ditched the old Julian calendar, with the consequence that 3–13 September 1752 never took place. The story that in London rioters objecting to the change chanted 'Give us our eleven days' may originate in a placard shown in *An Election Entertainment*, a satirical print by William Hogarth of a 1754 parliamentary election in Oxfordshire, during which the Tories attacked the Whigs on every conceivable front, including the calendar reform introduced by the latter. There does not appear to be any evidence that any 'calendar riots' actually took place.

PREDICTING THE DATE OF ONE'S OWN DEATH

(27 November) Death in exile in London of the French Huguenot mathematician Abraham de Moivre, fellow of the Royal Society, friend of Isaac Newton and author of *The Doctrine of Chances*, a book of great interest to gamblers. Having noted in old age that he was sleeping an additional quarter of an hour each day, he made an extrapolation and predicted that on 27 November 1754 he would sleep for a full 24 hours – and die. And so it came to pass.

AN UNTUTORED ARITHMETICAL PRODIGY

Jedediah Buxton, a farm labourer who had received so little education that he was unable to write, walked from his home in Elmton, Derbyshire, to London, where the Royal Society was anxious to test his reputation as a mental calculator. Buxton had, it was reported, measured out the Lordship of Elmton by counting his strides as he walked across it, and calculated its area in acres, square rods, square inches and even square 'hair's breadths', a hair's breadth being defined by him as $\frac{1}{48}$ of an inch. He also worked out how much you would have if you doubled a farthing 139 times; the result in pounds gave a number with 39 digits – a number that he then proceeded to multiply by itself.

Buxton was apparently unable to concentrate on anything other than numbers. While in London he was taken to see David Garrick play the title role in Shakespeare's *Richard III* at Drury Lane, and his attention was entirely taken up with counting the words spoken by the actor. He declared that 'the innumerable sounds produced by the musical instruments had perplexed him beyond measure'. The Royal Society were sufficiently impressed to award Buxton 'a handsome gratuity'. He died in 1772, and a portrait of him can be seen in Elmton Church.

HUMAN POPULATION IS CONSTANT

The great French *Encyclopédie* stated that 'Population is constant in size and will remain so right up to the end of mankind.'

ELECTRICAL STOCKINGS AND THE BAREFOOT PHILOSOPHER

Robert Symmer presented his findings on static electricity to the Royal Society. 'I had,' he said, 'for some time observed, that upon pulling off my stockings in an evening they frequently made a crackling or snapping noise,' and noted that no one had hitherto investigated this phenomenon. Symmer went on to conduct a number of experiments using stockings of different colours and materials, and abandoned the tiresome business of pulling the stockings on and off his legs, instead using his hands. He was aware that others were failing to take his work seriously:

> You may likewise be disgusted with the frequent mention of pulling on, and putting off, of stockings: a circumstance, I confess, so little philosophical, and so apt to excite ludicrous ideas, that I was not surprised to find it the occasion of many a joke, among a sarcastical set of minute philosophers, who do not love to have anything new forced on them.

The French, moderately impressed, nicknamed Symmer '*le philosophe déchaussé*' – the barefoot philosopher.

KILLED BY ADULTERATED WINE?

(14 April) Death of the composer George Frederic Handel. Over the course of a number of years he had suffered blindness, uneven gait, mood swings, garbled speech and several minor strokes and bouts of paralysis. In 2009 the American music historian David Hunter suggested that all these complaints may have been the result of lead poisoning. Handel famously quaffed large quantities of wine with his gargantuan meals, and since Roman times lead had routinely been used to sweeten sour wine.

More than two centuries after Handel's death, the sweet taste of lead almost did for an Australian building worker. In 1996 this man was admitted to hospital with stomach pains. It turned out he had been chewing electrical cables to dull his craving for cigarettes. The 'sweet and pleasant taste' that so appealed to him was due to the lead in the cables, added to give them greater flexibility. The man had chewed through a metre (3 ft) a day over a period of ten years, and the level of lead in his blood was three times the safe limit. Appropriate treatment restored him to good health – although he never managed to kick the nicotine habit.

AN AVERSION TO HUMAN CONTACT

The Hon. Henry Cavendish, in his paper 'Factitious Airs', announced his discovery of 'inflammable air' – later named 'hydrogen' by Antoine Lavoisier. Cavendish, who carried out pioneering experiments in many fields of science, was a painfully shy man, with a shrill voice (when he could bring himself to speak) and a shuffling gait. Lord Brougham said that he 'probably uttered fewer words in the course of his life than any man who ever lived to eighty years, not at all excepting the monks of La Trappe'. Cavendish's female servants received all their instructions in written form, and he was said to have built an extra staircase in his house so that he would not have to encounter his housekeeper. His social life was confined to meetings of the Royal Society, of which he was a fellow, but those who wished to ask him about his work found themselves speaking 'as if into a vacancy'. Despite his considerable inherited wealth, Cavendish always dressed plainly, and the only known portrait of him was sketched surreptitiously while he dined at the Royal Society Club. Cavendish died in 1810, having published very little, and it was not until the 1870s that the great physicist James Clerk Maxwell went through his papers and found he had anticipated many later discoveries in physics, such as Ohm's law and Charles's law of gases.

1767
Aeronautics

SLEIGHT OF HAND?

Following Cavendish's discovery of hydrogen the previous year, the Scottish chemist Joseph Black astonished his audience by pumping the gas into a sack, which then floated up to the ceiling. Sceptics accused him of mounting an elaborate hoax involving invisible threads. The first manned flight in a hydrogen-filled balloon took place on 1 December 1783.

CLAP AND POX ONE AND THE SAME, CLAIMS SURGEON

For centuries there had been confusion among physicians as to the differences and similarities of the two commonest sexually transmitted diseases, syphilis and gonorrhoea (known respectively as 'the pox' and 'the clap'). Hoping to clarify matters, the distinguished Scottish surgeon John Hunter inoculated himself with 'venereal matter' taken from a patient suffering from gonorrhoea, and over the next few months observed that he developed symptoms of both gonorrhoea and syphilis. This led him to conclude (wrongly) that the two were the same disease; in fact, the patient from whom he had taken the 'venereal matter' was more than likely suffering from both diseases. It was not until 1879 that the gonococcus, the microbe that causes gonorrhoea, was found, while the bacterium that causes syphilis was identified in 1905.

THE CHESS-PLAYING TURK

Baron Wolfgang von Kempelen, a Hungarian nobleman, began to tour his chess-playing automaton around the capitals of Europe, including Paris, Vienna and London. It consisted of a large box with a chess board on top, and a wooden figure dressed in Turkish clothes sat cross-legged on a chair attached to the box. This figure was operated by gears and wires apparently powered by clockwork inside the wooden box, and would nearly always beat anyone who challenged it to a game of chess – including Benjamin Franklin. In 1805, after the Baron's death, the machine was acquired by Johann Nepomuk Maelzel, who continued to tour it round Europe, and also took it to America, where it was seen by Edgar Allan Poe, who described it in detail in 1836 in the *Southern Literary Messenger*:

During the progress of the game, the figure now and then rolls its eyes, as if surveying the board, moves its head, and pronounces the word *echec* [check] when necessary. If a false move be made by his antagonist, he raps briskly on the box with the fingers of his right hand, shakes his head roughly, and replacing the piece falsely moved, in its former situation, assumes the next move himself. Upon beating the game, he waves his head with an air of triumph, looks round complacently upon the spectators . . .

Poe thought there must be a dwarf hidden inside the Turk's trunk. In fact, it was the box that hid the secret operator, a full-grown man, who used levers and gears to move the Turk's arm, which in turn moved the chess pieces. And the reason that the Turk always won – revealed the year after Poe's article was published – was that the Baron, and Maelzel after him, always employed chess masters as the secret operators within the box.

ELECTRICAL STIMULATION OF TURNIPS

Publication of Arthur Young's *A Six Months' Tour through the North of England*, in which he discusses the agricultural research of a Mr Clarke of Belford:

> An experiment he tried of the effect of electricity on vegetation deserves attention; he planted two turnips in two boxes, each containing 24 lb [10.9 kg] of earth: he kept them in the same exposure, and all circumstances the same to each, save that one was electrified twice a day, for two months, at the end of which time it was in full growth, the skin bursting, and weighed 9 lb [4.1 kg]. The other, at the end of four months, did not quite reach that weight: a strong proof that the electric fire had a remarkable power in promoting and quickening the vegetation.

ON THE INIQUITIES OF TEA

Publication of *The natural history of the tea tree with observations on its medical qualities, and effects of tea-drinking*, in which the fashionable London physician, Dr John Lettsom – known as 'Dr Wriggle' on account of his social ambitions – argued that tea drinking had pernicious consequences for society, being at once enervating and effeminizing. The Lord Chancellor, Thomas Erskine (1750–1823), penned the following epitaph for Dr Lettsom:

> Whenever patients come to I,
> I physics, bleeds and sweats 'em;
> If after that they choose to die,
> What's that to me – I lets 'em.

1774
Geodesy

FIDDLING ON THE FAIRY HILL

The Astronomer Royal, the Reverend Dr Nevil Maskelyne, travelled to Scotland on behalf of the Royal Society in order to conduct experiments on the Perthshire mountain called Schiehallion (whose name means 'fairy hill of the Caledonians'). The purpose was to 'weigh the world' – in other words, to determine the density of the Earth. Schiehallion was chosen because of its conical shape and its symmetry, and for four months Maskelyne and his assistants, living in a hut on the slopes of the mountain, conducted experiments involving observations of the deviation of plumb lines, and of stars near their zeniths on the north and south sides of the mountain. From Maskelyne's observations, Charles Hutton calculated that the density of the Earth was 4.5 times that of water (the currently accepted value is 5.515). Although the expedition was a success, Maskelyne appears to have derived little pleasure from the experience:

My going to Scotland was not a matter of choice, but of necessity. The Royal Society . . . made a point with me to go there to take the direction of the experiment, which I did, not without reluctance, nor from any wish to depart from my own observatory to live on a barren mountain, but purely to serve the Society and the public, for which I received no gratuity, and had only my expenses paid for me.

The Royal Society did not, however, cover the cost of the farewell party, organized by the expedition cook, Duncan Robertson, who laid on whisky and played the fiddle. Such was the exuberance of the occasion that the hut burnt down, and Robertson's fiddle also fell victim to the flames. In recompense, once he was back in London, Maskelyne sent Robertson a new fiddle – a Stradivarius.

ANIMAL MAGNETISM AND THE BIRTH OF HYPNOSIS

The Austrian physician Anton Mesmer conducted his first experiment in 'animal magnetism'. Mesmer held that just as the Sun and the Moon affect the tides, so they affect the free flow of the 'process of life' through the human body. To mimic this effect, he attempted to create an 'artificial tide' in a patient by first getting her to drink a potion containing iron, and then attaching magnets to various parts of her body. She reported that her symptoms disappeared for a few hours, as she felt the 'artificial tide' wash around within her. Mesmer himself believed that his own 'animal magnetism' played a greater part in her 'cure' than the magnets, and in subsequent treatments he no longer deployed them. He subsequently set up a successful practice in Paris, where, in 1784, Louis XVI set up a commission to study his methods. The commission, which numbered the chemist Antoine Lavoisier and the American ambassador Benjamin Franklin

A satirical print showing the fashionable Austrian physician Anton Mesmer treating his patients with what he claimed was 'animal magnetism'.

among its members, concluded that there was no mysterious fluid involved in Mesmer's cures, which they attributed entirely to 'imagination'. Mesmer's technique – which appeared to be highly effective in cases that we would now call psychosomatic or hysterical – became known as 'mesmerism'.

In 1841 the Scottish physician James Braid, after witnessing a demonstration by a mesmerist, sought to put Mesmer's therapeutic method on a more rational footing. He renamed it 'neuro-hypnotism', and stated that 'the phenomena are solely attributable to a peculiar physiological state of the brain and the spinal cord' induced by 'causing the patient to fix his thoughts and sight on an object, and suppress his respiration'.

Despite these rationalizations, the public continued to find mesmerism/hypnotism – with its ability to overwhelm the individual will – both mysterious and alarming. Edgar Allen Poe, in his 1845 story 'The Facts in the Case of M. Valdemar', took things one step further, for if mesmerism could suppress the individual will, perhaps it might also effect an individual's ultimate destiny? Thus in Poe's story a mesmerist describes how he succeeds in delaying the death of a terminally ill patient, Monsieur Ernest Valdemar. With only a few hours left to live (according to his doctors), Valdemar is put into a trance, and will only move at the instruction of the mesmerist. A day passes, and at last Valdemar's breathing ceases and his heart stops – and yet his brain, still controlled by the mesmerist, continues to prompt him to croak out answers to the questions put to him. It is only after seven months that the mesmerist takes Valdemar out of his trance:

> As I rapidly made the mesmeric passes, amid ejaculations of 'dead! dead!' absolutely *bursting* from the tongue and not from the lips of the sufferer, his whole frame at once – within the space of a single minute, or less, shrunk – crumbled – absolutely *rotted* away beneath my hands. Upon the bed, before the whole company, there lay a nearly liquid mass of loathsome – of detestable putrescence.

A MATHEMATICAL PROOF OF THE EXISTENCE OF GOD?

In his *Souvenirs de vingt ans de séjour à Berlin* (1804), a certain Monsieur Thiébault tells the story of Denis Diderot's visit in 1774 to the court of Catherine the Great in Russia. Diderot, an avowed atheist, was confronted by an unnamed Russian mathematician and philosopher, who addressed the French *philosophe* thus: '*Monsieur, $a + b^n/n = x$*. Therefore God exists. Your response, please!' Diderot, realizing that he was being subjected to ridicule, shortly afterwards returned to France. In some versions of this story, Diderot's nemesis is the much-celebrated Swiss mathematician Leonhard Euler, a man of simple faith but great intellectual ability, who in 1766 had been invited to live in Russia by Catherine the Great.

PEOPLE REMAIN UNCOOKED WHILE STEAK SIZZLES

The English physician Charles Blagden carried out a series of experiments – reported the following year to the Royal Society – on the effects of high temperatures on human beings. To this end he constructed what would now be called a sauna, and invited a number of persons to participate, including the eminent botanist Joseph Banks, who four years later was to become president of the Royal Society. At first the men sat around fully clothed, but as Blagden stoked the heat up to 100°C (210°F), and then to 127°C (260°F), clothes were abandoned. At such temperatures, Blagden found that steak was thoroughly cooked in 13 minutes, whereas the men emerged from the room without suffering any adverse effects. Blagden, puzzled that dead meat reacted differently to living flesh, concluded that nature had provided living organisms with some unique way of 'destroying heat'. At that time, it was not understood that evaporation has a cooling effect – this is the role of sweating – or that at higher temperatures the surface blood vessels expand, making them more effective radiators of heat.

ON THE DANGERS OF LIVING TOO FAST

Joseph Priestley published *Experiments and Observations on Different Kinds of Air*, describing the gases given off by heating various substances, and the effect of these different 'airs' upon mice. He was anxious that his experimental animals should suffer no more than necessary: 'It will be proper,' he wrote, '. . . to keep hold of their tails, that they may be withdrawn as soon as they begin to show signs of uneasiness.' Heating red mercuric oxide, he found that the gas given off made a candle burn 'with more splendour and heat', and that a mouse could live twice as long in a sealed container filled with this gas as it could in a container filled with ordinary air. He called the gas 'dephlogisticated air', and considered what its effects might be on humans:

> . . . though pure dephlogisticated air might be very useful as *medicine*, it might not be so proper for us in the usual healthy state of the body: for, as a candle burns out much faster in dephlogisticated than in common air, so we might, as may be said, *live out too fast*, and the animal powers be too soon exhausted in this pure kind of air. A moralist, at least, may say, that the air which nature has provided for us is as good as we deserve.

Priestley's 'dephlogisticated air' was, of course, oxygen, but it took Antoine Lavoisier to finally dispose of the phlogiston theory and give oxygen its name (*see* 1789).

1779
Rocket science
DARWIN'S GRANDFATHER DESIGNS ROCKET ENGINE

Erasmus Darwin, grandfather of the more famous Charles, designed a simple rocket engine powered by liquid fuel. In his sketch he showed separate tanks for oxygen and hydrogen, linked by pumps and pipes to an elongated combustion chamber with an expansion nozzle. It was not until 1926, however, that the first liquid-fuelled rocket – built by the American physicist Robert Goddard and powered by petroleum and liquid oxygen – was actually launched. The site of this historic event was Goddard's Aunt Effie's farm at Auburn, Massachusetts, and the rocket, christened *Nell*, rose 12.5 m (41 ft) into the air. After a flight lasting 2.5 seconds, *Nell* landed in a field of cabbages.

A DARK DAY IN NEW ENGLAND

(19 May) During the morning, the sky across New England and parts of Canada became so dark that birds began to roost, and animals became filled with terror. The members of the Connecticut legislature, fearing that the end of the world was come, moved for adjournment. But one member, Abraham Davenport, was made of sterner stuff, declaring: 'The day of judgement is either approaching, or it is not. If it is not, there is no cause of an adjournment. If it is, I choose to be found doing my duty. I wish therefore that candles may be brought.' By midnight, the skies had cleared, and the stars could once more be seen. Davenport and the 'Dark Day' became the subject of an 1866 poem by John Greenleaf Whittier:

> 'Twas on a May-day of the far old year
> Seventeen hundred eighty, that there fell
> Over the bloom and sweet life of the Spring,
> Over the fresh earth and the heaven of noon,
> A horror of great darkness, like the night
> In day of which the Norland sagas tell,
> The Twilight of the Gods. The low-hung sky
> Was black with ominous clouds, save where its rim
> Was fringed with a dull glow, like that which climbs
> The crater's sides from the red hell below.
> Birds ceased to sing, and all the barnyard fowls
> Roosted; the cattle at the pasture bars
> Lowed, and looked homeward; bats on leathern wings
> Flitted abroad; the sounds of labour died;
> Men prayed, and women wept; all ears grew sharp
> To hear the doom-blast of the trumpet shatter
> The black sky, that the dreadful face of Christ
> Might look from the rent clouds, not as He looked
> A loving guest at Bethany, but stern
> As Justice and inexorable Law.

It is thought that the darkening – which began at different times at different places – was caused by smoke from forest fires combining with fog and thick cloud cover. For some days before the Dark Day, the Sun had appeared red and the sky yellow, and after the darkness passed and true night came, the Moon was red as blood.

WIND TO BE RENDERED AS AGREEABLE AS PERFUMES

In his letter 'To the Royal Academy of Farting', Benjamin Franklin, much taken with the discomfort of holding in wind – necessary if one is to avoid giving offence in company – suggested a prize question for budding scientists:

> To discover some drug wholesome and not disagreeable, to be mixed with our common food, or sauces, that shall render the natural discharges of wind from our bodies not only inoffensive, but agreeable as perfumes.

ON THE EFFECTS OF ASPARAGUS

In the same letter, Franklin observed:

> A few stems of asparagus eaten, shall give our urine a disagreeable odour; and a pill of turpentine no bigger than a pea, shall bestow on it the pleasing smell of violets.

The odour in urine attributable to the consumption of asparagus – experienced by roughly half the population – is a result of the body breaking down a chemical in the asparagus (thought to be asparagusic acid), producing methyl mercaptan and various other sulphur-containing compounds. The ability to produce the smell is genetically determined.

Another vegetable that can have an interesting effect on the urine is beetroot (known to the Victorians as 'blood turnip'): after eating it, about one in ten people find that their urine turns red. This condition, known as beeturia, has led to a number of unnecessary visits to the doctor. Blackberries, which, like beetroot, possesses the pigment anthrocyanin, can have the same effect.

A SHEEP CALLED CLIMB-TO-THE-SKY

(18 May) The French astronomer Jérôme de Lalande wrote in the *Journal de Paris* that 'It is entirely impossible for man to rise into the air and float there. For this you would need wings of tremendous dimensions and they would have to be moved at three feet per second. Only a fool would expect such a thing to be realized.' Lalande had not taken into account the ingenuity of two of his compatriots, the brothers Joseph-Michel and Jacques-Étienne Mongolfier. Six months after Lalande's article appeared, Joseph Montgolfier, having noted how laundry drying over a fire was wafted upward, constructed a box-like chamber with a thin wooden frame and covered with taffeta cloth. Under the stand on which it stood, he set a small fire, and watched with satisfaction as the box rose up to the ceiling. Plans for a manned flight in a much larger contraption soon developed, but there was some concern that humans might find the upper atmosphere deleterious to their health. The king, Louis XVI, suggested sending up two criminals, but the first living creatures to take flight in the Montgolfiers' new hot-air balloon, on 19 September 1783, were a duck, a cockerel and a sheep called Montauciel (French for 'climb-to-the-sky'), who rose to an altitude of 460 m (1500 ft), travelled 3 km (2 miles) and landed safely after a flight lasting eight

Consternation ensues as a duck, a cockerel and a sheep called Montauciel safely land after the first 'manned' flight in a hot-air balloon.

minutes. The first manned flight took place on 15 October, when a young physician, Pilâtre de Rozier, rose up into the air in a tethered balloon 23 m (75 ft) high and 14 m (46 ft) in diameter.

THE FIRST GLIMMER OF A BLACK HOLE

The Reverend John Michell, a distinguished astronomer and the rector of Thornhill in Yorkshire, presented a paper to the Royal Society in which he suggested that there could be stars so massive that their gravitational attraction would prevent light from escaping from them. He thus anticipated not only Einstein's theory of general relativity, but also the concept of black holes, which modern cosmologists first hypothesized in the 1930s as a means of explaining what happens when large stars collapse at the end of their lives.

HIGH SPEEDS MAY CAUSE DEATH, OPINES DOCTOR

(August) John Palmer initiated an express mail coach service between Bath and London (a distance of around 100 miles) that cut the journey time from 38 to 16 hours. Some were alarmed by this new development. 'Regular travel at such a prodigious speed,' a concerned physician wrote to the *Bath Argus*, 'must surely result in death from an apoplexy.' Similar fears were voiced a generation or two later, with the coming of the railways. 'Rail travel at high speed is not possible,' opined Dionysus Lardner, professor of natural philosophy and astronomy at University College, London, in around 1830, 'because passengers, unable to breathe, would die of asphyxia.' A century later, the same old song was being sung. In his 1936 book *After Us*, John P. Lockhart-Mummery, a fellow of the Royal College of Surgeons, warned: 'The acceleration which must result from the use of rockets inevitably would damage the brain beyond repair.' Lockart-Mummery was closer to the mark than his predecessors: both fighter pilots and astronauts have to be carefully trained to safely withstand high G-forces.

FRENETIC FROGS AND CONVULSING CORPSES

The Italian physician and physicist Luigi Galvani was dissecting a frog at a table during a thunderstorm when he noticed that when the frog's sciatic nerve came into contact with a scalpel (which had picked up a charge) the frog's muscles twitched. Subsequently, he used brass hooks to hang a row of dissected frogs on an iron railing, and noticed that the muscles contracted when they touched the iron, even when there was no thunderstorm in progress. Galvani concluded the muscle movement was caused by an electrical fluid that flowed through the nerves to the muscles. It took his contemporary, Alessandro Volta, to show that what Galvani dubbed 'animal electricity' – believing it to be a uniquely animal phenomenon – was not any different from the current Volta generated in his 'voltaic piles', which consisted of a series of plates (electrodes) made from different metals separated by a saline solution (the electrolyte). In the case of the frogs, the brass hooks and iron railing had acted as the electrodes, and the fluids within the frogs' tissues as the electrolyte. Nevertheless, Galvani's concept of animal electricity and the description of his experiments influenced Mary Shelley's account of the creation of the monster in her novel *Frankenstein*, published in 1818, two decades after Galvani's death.

Whether the novelist was also familiar with the work of Galvani's nephew, the physicist Giovanni Aldini, is unknown. In pursuit of his uncle's ideas, Aldini put on public demonstrations in which he would pass electrical currents through humans and animals, producing twitches, jerks and spasms of the face and limbs. A spectator left an account of one such demonstration:

> Aldini, after having cut off the head of a dog, makes the current of a strong battery go through it: the mere contact triggers really terrible convulsions. The jaws open, the teeth chatter, the eyes roll in their sockets; and if reason did not stop the fired imagination, one would almost believe that the animal is suffering and alive again.

At the Royal College of Surgeons in 1803, Aldini experimented before a medical and general audience upon the body of a newly hanged criminal, George Forster, touching various parts of his body with rods attached to a powerful battery. When the rods touched the mouth and ear, 'the jaw began to quiver, the adjoining muscles were horribly contorted, and the left eye actually opened'. The most dramatic result occurred when a rod was applied to Mr Forster's rectum, which caused the entire body to convulse so much 'as almost to give an appearance of re-animation'. It was such experiments that gave us the verb 'to galvanize'.

1788
Zoology /
immunology

JENNER HONOURED FOR WORK ON CUCKOOS

Edward Jenner was elected a Fellow of the Royal Society for his work on elucidating the life of the infant cuckoo in the nest. Nine years later, when he sent the Society a paper reporting his pioneering experiments on vaccination, he was told that he 'ought not to risk his reputation by presenting to the learned body anything which appeared so much at variance with established knowledge, and withal so incredible'. What the Society had not foreseen was that by 1979 vaccination had wiped out smallpox from every single part of the world.

THE TRIUMPH OF OXYGEN OVER PHLOGISTON

To celebrate his demonstration of the role of oxygen in combustion, and the consequent confounding of the phlogiston theory of Georg-Ernst Stahl that had dominated scientific thinking for half a century, the French chemist Antoine Lavoisier, a vain and rapacious man, put on a masque at his house in Paris, to which he invited the *haute monde* of the city. In this entertainment the phlogiston theory, personified by a decrepit old man wearing a mask to resemble Stahl, was put on trial, the charge being read out by 'Oxygen', a personable young man. The accused was found guilty and the judges, among whom Lavoisier himself sat, passed sentence. Then Stahl's book was consigned to the flames by Mme Lavoisier, dressed in the white robes of a priestess. This was shortly after the Fall of the Bastille.

In 1794, during the Terror, Lavoisier himself was put on trial for his role as a 'tax farmer', and sent to the guillotine. There is a story, apparently apocryphal, that in order to see how long consciousness would be maintained after the fall of the blade, he determined to blink as many times as he could after his head was severed from his body, and ordered his manservant to record the result. Supposedly, there were between 15 and 20 blinks – although such manifestations (recorded in other cases) are more likely to have been muscle reflexes, as consciousness is probably lost within two or three seconds of decapitation.

WET SHIRTS IN THE PACIFIC

Having been cast adrift by the mutineers aboard the *Bounty*, Captain Bligh instructed his loyal companions with whom he shared a small open boat to keep their clothes soaked with sea water throughout their remarkable 47-day voyage across 6700 km (4100 miles) of the Pacific Ocean to Timor. The mutineers had only provided the castaways with food and water for a few days, and Bligh realized that they could only survive by catching fish and trapping rain water. The idea behind keeping their clothes constantly wet was that the cooling effect of the evaporation of the sea water would reduce the amount of fluids the men lost by sweating under the hot tropical sun. In 1951 the Royal Navy conducted experiments in the Straits of Johor at the tip of the Malay Peninsula and found that if men on life-rafts kept their clothes wet in this way, their survival time, if they had little or no fresh water, could be extended from two to three days to two to three weeks.

A POETICAL BIG CRUNCH

In his long botanical poem *The Economy of Vegetation*, Erasmus Darwin anticipated one modern theory regarding the future of the Universe. This holds that the expansion begun at the Big Bang will eventually run out of momentum, slow down and stop. There will then follow a Big Crunch, as gravity pulls back all matter to a single point. This in turn will be followed by another Big Bang, marking the birth of a new Universe. In his poem, Darwin apostrophizes the stars thus:

Flowers of the sky! ye too to age must yield,
Frail as your silken sisters of the field!
Star after star from Heaven's high arch shall rush,
Suns sink on suns, and systems systems crush,
Headlong, extinct, to one dark centre fall,
And Death and Night and Chaos mingle all!
– Till o'er the wreck, emerging from the storm,
Immortal Nature lifts her changeful form,
Mounts from her funeral pyre on wings of flame,
And soars and shines, another and the same.

One of the alternative modern scenarios holds that if there is insufficient matter to exert the gravitational pull necessary for the Big Crunch, the Universe will continue to expand, and eventually, through the process of entropy, decay into smaller and smaller particles. This will culminate in what is called the heat death of the Universe. In yet another scenario, observations that celestial objects towards the edge of the observable Universe appear to be accelerating away from us has suggested to some that there is an as-yet unidentified anti-gravitational 'dark energy' at work.

1792
Technology

THE FIRST SAFETY COFFIN

Prior to his death, Duke Ferdinand of Brunswick ordained that he should be buried in a coffin supplied with a window, an air tube and a lock on the inside, while in his shroud were to be placed two keys, one for the coffin and one for the door of the mausoleum in which he was interred. Subsequent ideas for safety coffins included that suggested by a Pastor Beck later in the same decade. This involved the provision of a tube leading from the coffin to the outside, so that the priest could check on a daily basis whether the expected smell of rotting flesh was emanating from the corpse; if not, the coffin was to be dug up and opened, in case the occupant still lived.

In 1822, a certain Dr Adolf Goldsmuth demonstrated the efficacy of his own trademarked safety coffin by spending several hours interred within his invention, during which period he consumed a meal of soup, sausages and beer.

GIANT-HEADED CREATURES ON THE SUN

In *Philosophical Transactions of the Royal Society*, Sir William Herschel, the distinguished astronomer who had discovered Uranus, stated that the Sun was no different in kind to the planets, all of which he believed were inhabited. 'Its similarity to other globes of the solar system,' he claimed, 'leads us to suppose that it is most probably inhabited . . . by beings whose organs are adapted to the peculiar circumstances of that vast globe.' Sunspots he interpreted as breaks in the fiery outer atmosphere of the Sun, through which one could see the cooler, inner atmosphere that shielded the inhabitants. These creatures, he calculated, must have enormous heads, otherwise they would explode.

EXQUISITE PAIN

The German scientist and geographer Alexander von Humboldt published a book on his experiments with electricity. In one of these experiments he sought to suppress the pain resulting from the extraction of a tooth by inserting an electrode into the cavity. It had the reverse of the desired effect, and Humboldt was convulsed in agony.

ODOUR OF DEAD RAT

The Chinese poet Shi Doanan reported that strange rats were emerging from the ground, spitting blood and then expiring. The rats were probably infected with plague, as shortly afterwards many people died, the disease-carrying fleas having abandoned their dying rodent hosts in favour of still-healthy humans. Shi Doanan believed the humans died from breathing odour of dead rat.

A PSEUDOSCIENCE IS BORN

The German anatomist and physiologist Franz Gall announced a new science, which he called 'cranioscopy', but which his followers were to name 'phrenology'. Gall assumed that different parts of the brain had different functions, and that the surface of the skull reflects the relative development of these regions. Gall identified 27 'brain organs', each matched by bumps on the surface. For example, the bump on the back of the head at the base of the skull was the 'Organ of Amativeness', involved with sexual arousal. Other 'organs' were associated with parental love, friendship, self-defence and courage, murderousness, cunning and cleverness, covetousness and ownership, pride, vanity and ambition, various aspects of memory, each of the five senses, humour, religion, metaphysics, and so on. His general idea that different functions are located in different parts of the brain was vindicated in 1861 when the French surgeon Paul Broca demonstrated the location of the brain's speech centre. However, Gall's assumptions regarding the particular functions of different areas, and his assertion that the surface of the skull accurately mapped these areas, both proved groundless. When he first began to lecture in Vienna, Gall was condemned by the Roman Catholic Church, and in 1805 he was deported by the Austrian government.

Mainstream scientists also largely condemned phrenology, but Gall's teachings struck a chord with a popular audience, and throughout the 19th century many turned to phrenology as a guide to the character (moral or otherwise) and even the likely fate of individuals.

AN UNNECESSARY HYPOTHESIS

Pierre-Simon Laplace published the first two volumes of his *Mécanique céleste* ('celestial mechanics'), in which he built on Newton's work and provided the mathematical tools for determining the movements and positions of all the bodies in the Solar System at any time – past, present or future. When Laplace went to see Napoleon to present him with a copy of his work, there ensued a somewhat strained exchange, as described by W.W. Rouse Ball in *A Short Account of the History of Mathematics* (4th edn, 1908):

> Someone had told Napoleon that the book contained no mention of the name of God; Napoleon, who was fond of putting embarrassing questions, received it with the remark, 'M. Laplace, they tell me you have written this large book on the system of the universe, and have never even mentioned its Creator.' Laplace, who, though the most supple of politicians, was as stiff as a martyr on every point of his philosophy, drew himself up and answered bluntly, '*Je n'avais pas besoin de cette hypothèse-là.*' [I had no need of that hypothesis.]

THE HOAX THAT WASN'T

George Shaw, keeper of the department of natural history at the British Museum, received a bizarre specimen from Captain John Hunter, the governor of New South Wales. The creature appeared to have the body of a large mole, the tail of a beaver and the bill of a duck. Shaw described the creature in the *Naturalist's Miscellany*, but declared that it was 'impossible not to entertain some doubts as to the genuine nature of the animal, and to surmise that there might have been practised some arts of deception in its structure'. Other naturalists concurred with him, suspecting that it was just one more of those 'monstrous impostures which the artful Chinese had so frequently practised on European adventurers'. But as more and more specimens arrived from Australia, it was realized that this was a genuine, if bizarre, creature – the platypus, a furry, milk-producing mammal with a duck-like bill and a venomous spur on the hind foot, and which further defies nature by laying eggs rather than giving birth to live young.

1799
Pharmacology

THE THRILLING EFFECTS OF NITROUS OXIDE

At the Pneumatic Institute in Bristol, the young Humphry Davy determined to try out on himself the effects of breathing various 'factitious airs'. A dose of carbon monoxide proved almost fatal – 'I seemed sinking into annihilation,' he recorded, noting his pulse was 'threadlike and beating with excessive quickness'. A week later he inhaled 'carbonic acid' (possibly vaporized phenol); this scalded his epiglottis so badly that he nearly choked. He then turned his attention to the 'phlogisticated nitrous air' that had first been synthesized in 1775 by Joseph Priestley, who regarded the gas – what we now know as 'nitrous oxide' – as lethal. Undeterred, Davy inhaled four quarts (4.4 litres), and experienced 'highly pleasurable thrilling, particularly in the chest and extremities'. He also noted that the objects around him became 'dazzling', and that his hearing became more acute. Davy repeated the experiment again and again through the spring and summer, for the overriding sensation of breathing the gas was one of

EXHIBITION OF THE LAUGHING GAS.

THE Nitrous Oxide, or Laughing Gas, was discovered by Dr. Priestly, who produced it by abstracting a part of the Oxygen from the Nitric Oxide. It is composed of equivalent parts of Oxygen and Nitrogen. Before the time of Sir Humphry Davy, it was considered irrespirable: but by some very interesting experiments, he proved this opinion to be incorrect; he also wrote a work, entitled, "Researches on the Nitrous Oxide." It is named Laughing Gas on account of the very exhilarating emotions produced in those who respire it for a short time: laughing, dancing, jumping, acting, reciting, and (last but not least) fighting are amongst the prominent effects displayed by persons under its influence. The Febrile Miasma depresses and terrifies the mind as much as the Nitrous Oxide raises and enlivens it. The easiest way of making it is to dissolve Crystals of the Nitrate of Ammonia in a retort, over a strong flame; after the atmospheric air has passed away, the Gas will be given off in great abundance, and may be collected in bladders, or a gasometer, for use. Sulphur, Phosphorus, red hot Charcoal, or a Taper, will burn with great brilliance when immersed in Nitrous Oxide.

Engraved and Printed at the Exhibition - - H. & A. HILL, Printers, Castle Green, Bristol.

A 19th-century engraving illustrating the effects of breathing nitrous oxide, accompanied by instructions as to how to prepare the gas.

pleasure: 'Sometimes I manifested my pleasure by stamping or laughing only; at other times, by dancing around the room and vociferating.' His spirits, he said, were kept in a constant glow. Davy went on to introduce the gas to the young literary lions of the day, including the poets Coleridge and Southey. A witness described the varied effects it had on its subjects: 'One it made dance, another laugh, while a third, in his state of excitement, being pugnaciously inclined, struck Mr Davy rather violently with his fist.' A 'courageous young lady' was induced to try a breath or two and turned into a 'temporary maniac', running out of the house and down the street, in the course of which she 'leaped over a great dog in her way'. Although Davy and his friends noted that what they called 'laughing gas' dulled the sensation of pain, it was not until the 1840s that nitrous oxide came into use as an anaesthetic (*see* 1844).

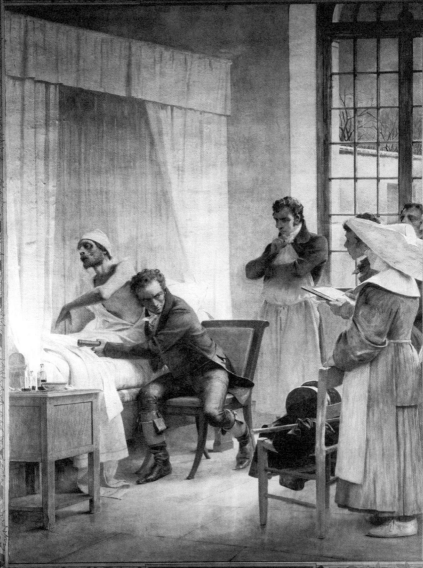

LAENNEC
· 1781 – 1826 ·

LAENNEC A L'HOPITAL NECKER
AVSCVLTE VN PHTYSIQVE DEVANT SES
ELEVES . 1816 .

1800 to 1849

War over bird droppings * How
the giraffe got its long neck *
Embarrassment the mother of invention
* Bathing one's testicles in cold water *
Enhancing the art of the cocktail * The
mystery of Napoleon's penis * Medicine's
loss, music's gain * Gastric juices and
gunshot wounds * On the amorous
effects of turbots and truffles * Eating
black vomit in the name of science *
Creatures on Moon evidently engaged
in conversation * The lost formula *
Abolition of pain a chimera * Birth of
the stock cube * A human calculating
machine * Balloon hoax punctured *
Crucifying a corpse * Unspeakable
disgust and pain in the service of science

WAR OVER BIRD DROPPINGS

The German explorer and geographer Alexander von Humboldt found that the substance the Peruvians spread on their fields was some 30 times richer as a fertilizer than European farmyard manure. This was *guano*, the Quichua word for 'the excreta of seabirds'. Guano – particularly that produced by the promisingly named Guanay cormorant – is rich in both phosphorus and nitrates, making it both an important input in agriculture and a key ingredient in explosives. The War of the Pacific (1879–83) between Chile and the alliance of Peru and Bolivia was largely fought over the control of this valuable substance.

HOW THE GIRAFFE GOT ITS LONG NECK

Jean-Baptiste Lamarck published *Zoological Philosophy*, in which he outlined his theory of evolution via the inheritance of acquired characteristics. According to Lamarck, giraffes acquired their long necks via the efforts of their ancestors to stretch up to eat the highest leaves:

> . . . this animal . . . is known to live in the interior of Africa in places where the soil is nearly always arid and barren, so that it is obliged to browse on the leaves of trees and to make constant efforts to reach them. From this habit long maintained in all its race, it has resulted that the animal's fore-legs have become longer than its hind legs, and that its neck is lengthened to such a degree that the giraffe, without standing on its hind legs, attains a height of six metres [20 feet].

This improvement of the species by an effort of will appealed to many (including George Bernard Shaw), but Lamarck's theory was demolished when Darwin showed that evolution took place via the blind and purposeless processes of natural selection.

A PERPETUAL PRANK

In Philadelphia, a man called Charles Redheffer charged the public one dollar a head to view what he described as a perpetual-motion machine. It was a sensational success, and Redheffer requested funds from the city to build a larger version. However, when eight city commissioners came to inspect his machine in January 1813, they noticed that the cogs on the gears via which the machine supposedly powered another device were worn on the wrong side, indicating that this other device was actually powering the machine. After Redheffer was exposed as a fraud, he moved to New York City, where he demonstrated another 'perpetual-motion machine'. When the engineer Robert Fulton (builder of one of the first successful steamboats) came to inspect it, he noticed irregularities in the sound and movement of the machine. Looking more closely, he found a hidden cord leading to the floor above – where an old man was turning a crank while eating a hunk of bread. Redheffer left the city in some haste. *See also* 1150.

EMBARRASSMENT THE MOTHER OF INVENTION

Confronted with the need to listen to the heart of a well-endowed young lady, the French physician René Théophile Hyacinthe Laënnec shrank from the indelicacy of placing his head upon her chest. Instead, he rolled up his notebook, placed one end of the tube on the bosom in question, and the other to his ear. He could in this manner hear all her inner workings quite satisfactorily – and thus was born the stethoscope.

BATHING ONE'S TESTICLES IN COLD WATER

Death of Samuel Solomon, whose universal panacea, which he called the Cordial Balm of Gilead, proved enormously popular as a cure for 'nerves'. He also recommended it for the treatment of 'impotency and seminal weakness', which he listed among the debilities 'arising from self-abuse'. Solomon did not claim that the Cordial could work on its own, however, and he advised patients to bathe their testicles in cold water or a combination of vinegar and alcohol. It turned out that the principal ingredient of his Cordial was brandy, with a dash of cardamom and cantharides.

ENHANCING THE ART OF THE COCKTAIL

Johann Gottlieb Benjamin Siegert, surgeon-general in Simon Bolivar's army of liberation in Venezuela, developed a medicine from the angostura tree to improve the appetite and aid the digestion of the troops. He thus inadvertently invented angostura bitters, the crucial ingredient in pink gins, Manhattans and a variety of other cocktails.

THE MYSTERY OF NAPOLEON'S PENIS

1821
Pathology

(6 May) The day after Napoleon's death on St Helena, his body was subjected to an autopsy, in the presence of 17 witnesses, including a priest called Vignali. The emperor's doctor, Francesco Antommarchi, removed Napoleon's heart, which the great man had requested be forwarded to his estranged wife, Empress Marie-Louise, although she never received it. Quite what else was removed remains unclear, but in 1916 descendants of Vignali's family sold his collection of Napoliana, including 'the mummified tendon taken from Napoleon's body during the post-mortem'. This subsequently came to be taken for the emperor's penis, although those who have seen it compare it to a shrivelled eel or an inch-long grape. The collection changed hands several times during the 20th century, and the putative penis is now thought to be in the hands of an American urologist.

MEDICINE'S LOSS, MUSIC'S GAIN

1821
Medicine

The young Hector Berlioz, in accordance with his father's wishes, began his medical studies in Paris. Nothing had prepared him for his first visit to the dissecting room in the Hospice de la Pitié:

> At the sight of that terrible charnel-house – the fragments of limbs, the grinning faces and gaping skulls, the bloody quagmire underfoot and the atrocious smell it gave off, the swarms of sparrows wrangling over scraps of lung, the rats in their corner gnawing the bleeding vertebrae – such a feeling of revulsion possessed me that I leapt through the window of the dissecting room and fled for home as though Death and all his hideous train were at my heels.

Although he soon hardened himself to the sights and smells of the dissecting room – even feeding the sparrows titbits of lung while tossing a shoulder blade 'to a great rat staring at me with famished eyes' – it was not long before he abandoned medicine to concentrate on his first love: musical composition.

DISASTER AVERTED

Charles Daubeny was appointed to the Aldrichian chair of chemistry at Oxford University. During one lecture, he held up two flasks and declared that if the liquids inside the flasks should mix, the entire lecture theatre would be destroyed in a vast explosion. He then turned round and tripped over, dropping the two bottles onto the floor. The audience gasped, shrank back, then breathed a sigh of relief. A thoughtful technician had seen fit to fill the flasks with harmless substances prior to the lecture.

GASTRIC JUICES AND GUNSHOT WOUNDS

A soldier called Alexis St Martin, stationed at Fort Mackinac on the US–Canadian border between Lakes Michigan and Huron, was badly injured in the stomach by the accidental discharge of a shotgun. An American military surgeon called William Beaumont was summoned, but did not hold out much hope for the wounded man. However, drawing on all his experience in treating gunshot wounds, Beaumont extracted bits of bone, shot and clothing from the wound, and applied a poultice. Despite running a high fever, St Martin pulled through – but the wound did not completely heal, and a hole the size of a man's fist (a 'fistula') led from the exterior directly into his stomach. Spotting a marvellous opportunity for scientific research,

Beaumont kept St Martin under his care for some years, and on 1 August 1825 inserted pieces of meat, bread and cabbage, secured by a silk thread, into St Martin's stomach via the fistula. At various intervals, Beaumont withdrew the pieces of food, to observe the progress of digestion. In other experiments he extracted some of St Martin's gastric juice, and dropped pieces of meat into it to observe what would happen. From this he made the important discovery that digestion is primarily a chemical rather than a mechanical process. Over the following years, Beaumont conducted further experiments on his not-always-willing subject, and published his findings in 1833. St Martin long outlived Beaumont: the latter died in 1853, while the man with a hole in his stomach lived on until 1880.

1825
Physiology

ON THE AMOROUS EFFECTS OF TURBOTS AND TRUFFLES

The French gastronomic writer Anthelme Brillat-Savarin, in his *Physiologie du gout* ('physiology of taste'), stated:

> Opinion is unanimous that fish are strongly sexual and awaken in both sexes the instinct for reproduction . . . These physical truths were without doubt unknown to the ecclesiastical law makers who imposed a Lenten diet on various priestly orders . . . for it is impossible to believe they could deliberately have wished to make even more difficult that vow of chastity already so antisocial in its observances.

Brillat-Savarin was equally sure of the aphrodisiac effects of truffles:

> Whosoever pronounces the word *truffle* . . . awakens erotic and gastronomical dreams equally in that sex that wears skirts and the one that sprouts a beard.

Le Problème du gros Turbot

Anthelme Brillat-Savarin demonstrates his method of cooking turbot. The consumption of any fish, he declared, awakens 'in both sexes the instinct for reproduction'.

An old French proverb has it that 'Those who desire to follow a virtuous path needs must leave truffles well alone.' There may be some truth in this. Truffles contain the hormone androstenol, and German scientists have found that if men and women are sprayed with a dilute solution of this hormone they become sexually aroused.

1828
Aetiology

EATING BLACK VOMIT IN THE NAME OF SCIENCE

Dr Nicholas Chervin of Gibraltar decided to test his theory that yellow fever was not transmissible from patient to patient. To this end he donned the soiled linen of those who had died of the disease, slept in their beds – and even ate the 'bloody black vomit' that victims produce in such spectacular quantities. Although Chervin survived the experiment, it was not until many decades later that it was finally proved that yellow fever was transmitted not by human contact but by mosquitoes.

ON THE NUTRITIONAL VALUE OF SAWDUST AND MUD

In his *Preliminary Discourse on the Study of Natural Philosophy*, the mathematician and astronomer Sir John Herschel wrote:

> Sawdust . . . is susceptible to conversion into a substance bearing no remote analogy to bread; and though certainly less palatable than that of flour, yet no way disagreeable, and both wholesome and digestible as well as highly nutritive . . . This discovery, which renders famine next to impossible, deserves a higher degree of celebrity than it has obtained.

In fact wood, being some 50 per cent cellulose, is impossible for humans to digest. Only creatures such as termites can derive energy from it, with the aid of specialized symbiotic microbes in their guts.

In a similar vein, in 1957 Robert Beauchamp, director of the East African Fisheries Research Organization based in Jinja, Uganda, proposed that mud taken from the bottom of Lake Victoria, if dried and powdered, could be used as a feed for pigs and poultry. Analysis of the mud had shown that it was rich in nutrients, as owing to the low sulphur levels in the lake, dead animals and plants were slow to decompose and release their nutrients back into the water. Beauchamp himself declared that the mud was perfectly pleasant to eat, and fed it to his friends and family.

COMING AND GOING WITH THE COMET

(30 November) Birth of Mark Twain, two weeks after the perihelion of Halley's Comet. In his *Autobiography*, Twain wrote:

> I came in with Halley's Comet in 1835. It's coming again next year [1910], and I expect to go out with it. The Almighty has said no doubt, 'Now here are these two unaccountable freaks; they came in together, they must go out together.'

The next perihelion of the comet occurred on 20 April 1910. Twain died the following day.

1835
Astronomy

NEVER SAY NEVER

The French philosopher Auguste Comte, founder of positivism, concluded that there were limits as to what science could discover. For example, we would never establish the chemical composition of the stars, he said, as they are too far away for us ever to collect samples. Less than quarter of a century later, the German physicists Robert Bunsen and Gustav Kirchoff used the spectroscope they had built to show that one of the wavelengths of light emitted by the Sun coincided with the wavelength of light emitted by burning sodium. Following this breakthrough, the other elements in the Sun's spectrum were gradually identified. It was through spectral analysis of sunlight that in 1868 the element helium was first discovered – which is why it was so named, after the Greek word *helios*, meaning 'Sun'. Only in 1895 was helium discovered on Earth – a rather ironic refutation of Comte's gloomy prediction.

CREATURES ON MOON EVIDENTLY ENGAGED IN CONVERSATION

(August) In the *New York Times* a series of reports appeared, apparently from Sir John Herschel's astronomical expedition to the Cape of Good Hope. Herschel had, it seemed, made a number of thrilling discoveries regarding life on the Moon. Not only had he classified 'thirty-eight species of forest trees and nearly twice this number of plants', but also 'nine species of mammalian and five of oviparia'. Most exciting of all were the 'flocks of large winged creatures' somewhat resembling human beings. 'These creatures were evidently engaged in conversation . . . We hence inferred that they were rational beings.' The 'reports' turned out to be fabrications perpetrated by one of the newspaper's own journalists.

THE LOST FORMULA

(10 June) Death of André-Marie Ampère, the French physicist considered one of the pioneers of the new field of electromagnetism. It was his habit, when an inspiration struck him, to jot down his calculations on any handy surface with the piece of chalk he always had on him. On one occasion, while walking the streets of Paris, he was struck by a thought, and proceeded to scribble down a sequence of equations on the back of a horse-drawn cab – only to find his blackboard start off at speed, taking with it forever the solution he had long sought.

FRESH FIELDS OF CONQUEST

Sir John Eric Erichsen, later surgeon-extraordinary to Queen Victoria, declared that 'There cannot always be fresh fields of conquest by the knife; there must be portions of the human frame that will ever remain sacred from its intrusions, at least in the surgeon's hands. That we have already, if not quite, reached these final limits, there can be little question. The abdomen, the chest and the brain will be forever shut from the intrusion of the wise and humane surgeon.'

ABOLITION OF PAIN A CHIMERA

Professor Alfred Velpeau of the medical faculty at Paris University declared that 'The abolition of pain in surgery is a chimera. It is absurd to go on seeking it . . . Knife and pain are two words in surgery that must forever be associated in the consciousness of the patient. To this compulsory combination we shall have to adjust ourselves.' The following decade, the 1840s, saw the introduction of ether, chloroform and nitrous oxide as general anaesthetics in surgery (*see* 1844).

BIRTH OF THE STOCK CUBE

Justus von Liebig, one of the greatest chemists of the 19th century, developed a financially rewarding sideline when he developed a concentrated beef extract. He founded the Liebig Extract of Meat Company, and his invention subsequently became known by a familiar brand name: Oxo.

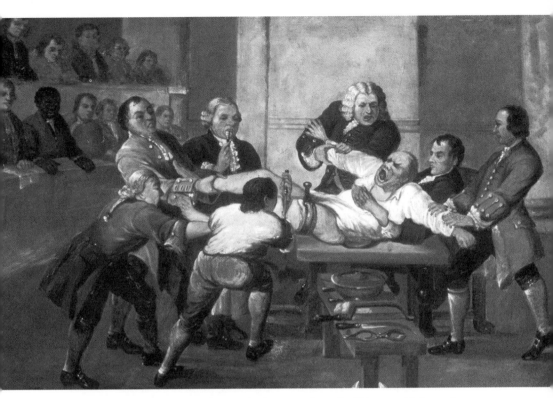

Surgery before the advent of anaesthesia was a brutal affair, even if the patient was rendered part-senseless by the administration of brandy.

1844
Anaesthesia

NO LAUGHING MATTER

It was Horace Wells, an American dentist, who first had the idea of using nitrous oxide (laughing gas; *see* 1799) as an anaesthetic. In 1844 he had attended a 'laughing gas frolic' – a party where the guests whiffed nitrous oxide for fun – and witnessed a young man stumble against a bench and break both legs. But the man felt no pain. Wells began to use nitrous oxide in his own practice, but when he mounted a demonstration of 'painless tooth extraction' to medical students at the Massachusetts General Hospital in 1845 he didn't use enough gas and the patient yelled out in agony. The students booed, and left the lecture theatre shouting 'Humbug!' Embittered and disgraced, Wells abandoned dentistry and became a travelling salesman. But his interest in anaesthesia never left him, and he embarked on extensive self-experimentation not only with nitrous oxide but also with ether and chloroform (although the 'sensations of exhilaration' he recorded were part of the appeal). The cumulative effects had a deleterious effect on his personality, and in January 1848 he was arrested after running into the street and throwing sulphuric acid at two women. He was committed to New York's Tombs Prison, where, coming to his senses and filled with remorse, he slit the femoral artery in his left thigh with a razor, inhaled some chloroform, and bled to death. In an anguished note to his family he expressed fears for his sanity, saying 'Did I live I should become a maniac.' His suicide coincided with his acclamation by the Medical Society of Paris as the discoverer of anaesthetic gases. But even this honour was to be taken away from him: in 1921 the American College of Surgeons declared that the first to use anaesthesia in surgery was not Wells, but the American physician Crawford Long, who in 1842 had used ether while excising tumours from a patient's neck and amputating a boy's toe.

1844
Mathematics
A HUMAN CALCULATING MACHINE

The German mental calculating prodigy Johann Martin Zacharias Dase worked out pi (*see* 1900 BC) to 200 decimal places in his head. Dase, who suffered from epilepsy, began a show-business career at the age of 15, travelling as far as England to demonstrate his abilities – which included multiplying 79532853 × 93758479 in 54 seconds. He could multiply two 40-digit numbers in 40 minutes, and but it took him 8 hours 45 minutes to multiply two 100-digit numbers. Dase attributed his astonishing skills to spending much of his boyhood playing dominoes.

BALLOON HOAX PUNCTURED

(13 April) The *New York Sun* carried the story of an extraordinary human achievement. 'The air, as well as the earth and the ocean, has been subdued by science,' it enthused, 'and will become a common and convenient highway for mankind.' The reason for this outpouring? 'The Atlantic has been actually crossed in a balloon . . . and in the inconceivably brief period of 75 hours from shore to shore!' Two days later the newspaper was obliged to print a clarification: 'The mails from the South last Saturday night not having brought a confirmation of the arrival of the Balloon from England . . . we are inclined to believe that the intelligence is erroneous.' In fact, the whole thing was a hoax concocted by Edgar Allan Poe, who forged the diaries of the supposed aeronauts, 'Mr Monck Mason and Mr Harrison Ainsworth', in which they describe the use of a propeller to achieve speeds of 'not less, certainly, than 50 or 60 miles an hour' (80–96 kph). There was just enough truth in the hoax for the newspaper to bite: Thomas Monck Mason (1803–89) was indeed a celebrated balloonist (and also a flautist and theologian), who in 1836 had published an account of an 18-hour balloon flight from London to Weilburg in Germany, a distance of 800 km (500 miles). Harrison Ainsworth (1805–82) was the author of *Rookwood*, *Jack Sheppard*, *The Tower of London*, *Old St Paul's* and many other historical novels. The first east-to-west Atlantic crossing by any aircraft was made by the British airship R-34 in an 108-hour flight in July 1919.

This was not the last of Poe's hoaxes. In the 14 April 1849 issue of *Flag of Our Union*, a popular weekly newspaper in Boston, he published 'Von Kempeln and His Discovery', a piece of apparent reportage in which he stated that Von Kempeln, a scientist of German extraction born in Utica, New York, had pursued a line of research embarked upon (he said) by Sir Humphry Davy, and recorded by Davy in his diary, which he had intended to be burnt after his death. Von Kempeln – whose family, Poe, wrote, 'is connected, in some way, with Maelzel' (*see* 1769) – had, it seems, found a method long sought by the alchemists: he had succeeded in transmuting base metal into gold. Poe proceeded to suggest that those thinking of joining the gold rush to California then in full swing might care to reconsider, given that once Von Kempeln's secret was out, gold would be no more valuable than lead – the price of which had already soared 'two hundred per cent' in Europe.

A MUSICAL EXPERIMENT

The Dutch scientist Christophorus Henricus Diedericus Buys Ballot – better known for his work in meteorology – sought to test the Doppler effect by placing an entire orchestra on a flat-bed railway truck on the line between Utrecht and Amsterdam, having instructed them to play a single note as they clattered past him to see if it appeared to change pitch. It did.

CRUCIFYING A CORPSE

(30 January) Death of the English surgeon and anatomist Joseph Constantine Carpue, famous for his reconstructive surgery on noses. Early in his career he had acquired some notoriety by crucifying a recently executed murderer on a cross to ascertain how the corpse would hang. He made a cast of the result, and published his findings in the *Lancet*. Many believed he was in league with the body snatchers (*see* 1752), and he was satirized in Thomas Hood's poem 'Mary's Ghost', in which Mary, having been disinterred, complains:

> I can't tell where my head has gone,
> But Dr Carpue can:
> As for my trunk, it's all packed up,
> To go by Pickford's van.

IF YOU CAN'T STAND THE HEAT . . .

One day the German-Swiss chemist Christian Friedrich Schönbein was experimenting at home in his kitchen, the laboratory at the University of Basel being closed for lunch. As he worked, he spilt some concentrated nitric acid on the kitchen table. Grabbing for the nearest cloth to hand, he wiped up the spillage with a cotton apron, then hung it up to dry above the stove. After a while there was a flash, and the apron disappeared. Thus was discovered a new, smokeless explosive: gun cotton, or nitrocellulose.

THE PLANET OCEANUS

In Berlin the German astronomer Johann Galle observed a new planet, following predictions as to its location sent to him by Urbain Le Verrier of the École Polytechnique in Paris. (In England John Couch Adams had independently predicted its existence.) Galle wanted to call the planet Janus, the two-faced Roman god of beginnings, while Le Verrier agitated for it to be named after himself – a suggestion popular in France, but not elsewhere. In Britain the astronomer royal, Sir George Airy, and the director of the Cambridge Observatory, James Challis, insisted it be named Oceanus, to be consistent with Uranus, its neighbouring planet discovered in 1781: in Greek mythology, the Titan Oceanus, the personification of the world ocean, is the son of Uranus and Gaia. However, in the end the new planet was called Neptune, the Roman god of the sea.

UNSPEAKABLE DISGUST AND PAIN IN THE SERVICE OF SCIENCE

(17 October) In a letter to Leonard Jenyns, Charles Darwin recollected his insect-collecting days while at Cambridge (1828–31), and in particular a dilemma presented by the discovery of three rare beetles:

> I must tell you what happened to me on the banks of the Cam in my early entomological days; under a piece of bark I found two carabi (I forget which) & caught one in each hand, when lo & behold I saw a sacred *Panagæus crux major*; I could not bear to give up either of my Carabi, & to lose *Panagæus* was out of the question, so that in despair I gently seized one of the carabi between my teeth, when to my unspeakable disgust & pain the little inconsiderate beast squirted his acid down my throat & I lost both Carabi & *Panagæus*!

The indirect cause of Darwin's discomfort, the Crucifix Ground Beetle (*Panagæus cruxmajor*), was thought to have been absent from Cambridgeshire for more than 50 years, until a specimen was rediscovered at Wicken Fen in 2008.

THE METAMORPHOSIS OF CLASSES TO LASSES TO ASSES

The renowned physicist William Thomson, later Lord Kelvin, was appointed to the chair of natural philosophy at the University of Glasgow at the age of just 22. One day he left a notice on the door of the lecture theatre that read: 'Professor Thomson will not meet his classes today.' Some wag rubbed out the 'c' in 'classes', putting a somewhat different spin on matters. Thomson was not slow in getting his own back. The next day, when his students turned up at the lecture theatre, they found that the professor had removed another letter, so that the notice now read: 'Professor Thomson will not meet his asses today.'

DESPERATE REMEDIES

Death of the Scottish surgeon Robert Liston. In the age before anaesthesia his speed with the scalpel and the bone-saw were highly valued, and he would encourage his watching students to time him on their pocket watches. On one occasion he amputated a man's leg in a two and a half minutes, although he inadvertently also removed the man's testicles at the same time. The patient later died of hospital gangrene, as did Liston's assistant, who lost his fingers to the surgeon's saw during the same frenzied operation. To cap it all, Liston in his haste somehow slashed the coat of a fellow surgeon who was observing the proceedings, and who, convinced that he had been stabbed, promptly died of fright.

1850 to 1899

A case of spontaneous human combustion? ∗ A nasty dish of soup ∗ The humanitarian argument for poison gas ∗ God plays tricks ∗ The Great Stink ∗ Linoleum prevents disease ∗ Dispensing with the necessity of stuffing ∗ Operating on the Queen's armpit ∗ A fog as dense as iron ∗ Freezing a man to death on a warm summer's day ∗ Doubts on the usefulness of the telephone ∗ Incandescent beards ∗ Darwin's soul absorbed with worms ∗ A case of penis captivus ∗ Supernatural instruction on the value of pi ∗ Ballybunion points the way to the future ∗ Cannabis a palliative for moral shock ∗ X-ray-proof underwear

DR MERRYWEATHER'S TEMPEST PROGNOSTICATOR

1850
Meteorology

Dr George Merryweather, honorary curator of the museum of the Whitby Literary and Philosophical Society, developed his Tempest Prognosticator or Atmospheric Telegraph Conducted by Animal Instinct. Inspired by the apparent reaction of medicinal leeches to changing electrical conditions in the atmosphere, Merryweather determined to use the creatures to forecast the weather. In his device, a number of leeches were placed in small bottles, and, when a storm approached, they would climb up the glass and so touch a strip of whalebone, which in turn would ring a bell. When it was pointed out to him that the leeches on occasion gave false warnings, he responded that there had indeed been a storm – in Denmark, or Dublin, or perhaps Dubrovnik. At the Great Exhibition in 1851 Merryweather showed a top-of-the-range version of his Tempest Prognosticator, in which the leeches were housed in what appeared to be an Indian temple.

A CASE OF SPONTANEOUS HUMAN COMBUSTION?

1850
Pathology

In its article on 'Spontaneous Combustion of the Human Body', *Chambers's Encyclopaedia of Universal Knowledge* (1888) mentions the following case, in which the renowned German chemist, Justus von Liebig, was called in as an expert witness:

> On the 13th of June 1847, the Countess of Goerlitz was found dead in her bedroom, with the upper part of her body partly consumed by fire. The head was a nearly shapeless black mass, with the charred tongue protruding from it. The physician who was consulted could suggest no

ALARMING CASE OF SPONTANEOUS COMBUSTION.

"OH! LAW! THERE'S PA'S BOOTS—BUT WHERE'S PA?"

'Pa', on leaving a gin palace, spontaneously combusts – alluding to the mistaken belief that humans, if sufficiently soused with alcohol, could suffer this fate.

other explanation than that the body of the countess must have taken fire spontaneously, and not even by ignition of her dress by a candle. On this evidence, she was buried; but circumstances having led to the suspicion that she had been murdered by her valet Stauff (who had been detected in attempting to poison the count), her body was exhumed in August 1848, fourteen months after her death, and was subjected to a special examination by the Hesse Medical College, who reported that she had not died from spontaneous combustion. The case was then referred to Liebig and [the pathological anatomist Theodor von] Bischoff, and their report was issued in March 1850, when Stauff was put upon his trial. They found no difficulty in concluding that the body was wilfully burned *after death*, for the purpose of concealing the murder (either by strangulation or a blow on the head), which had been previously perpetrated. The prisoner was convicted, and subsequently confessed that he had committed the murder by strangulation, as indeed the protruded tongue might have suggested. Since that date, there has not been any case of alleged spontaneous combustion.

In his *Letters on Chemistry* (3rd edn, 1851), Liebig dismisses those writers who assert that if a person is excessively fat, and filled with spirituous liquor, they are rendered abnormally flammable. He counters their assertions with 'the fact that hundreds of fat, well-fed brandy-drinkers do not burn, when by accident or design they come too near a fire'. As long as the circulation continues, he adds, their bodies could not ignite, and concludes that spontaneous combustion in a living body is absolutely impossible.

1851
Pharmacology

DOCTOR ADVOCATES BHANG

Publication of *The Male Generative Organs in Health and Disease, From Infancy to Old Age: being a complete practical treatise . . . adapted for every man's own private use* by the American physician Dr Frederick Hollick. In this he recommended a substance that induced feelings of warmth and cheer, left no depressing after effect, and which, most importantly, was effective in restoring potency and desire: cannabis.

COLOURING THE MAP

While colouring a county map of England, Francis Guthrie proposed that in cartography one only needs four colours so that no adjacent counties or countries will have the same colour. His conjecture was eventually proved – by computer – in 1976.

A NASTY DISH OF SOUP

In Dresden, Dr Friedrich Küchenmeister conducted a gruesome experiment to test his hypothesis that the presence of tapeworms in the intestines of humans was due to the consumption of infected meat. His suspicions had been aroused by the fact that pork butchers and their families appeared to suffer more than most people from these parasites. However, the 'bladder worms' that appeared in pigs (and also cattle) looked like little buttons with fringes of hooks – different in appearance to the long, thin tapeworms that infest humans. Some contemporary scientists held that these 'bladder worms' were deformed versions of the human tapeworm, which had ended up in the wrong host. Küchenmeister indignantly declared that this would be 'contrary to the wise arrangement of Nature', and to prove his hypothesis that the two forms were simply different stages in the same creature's life cycle, he persuaded the authorities to allow him to feed a soup containing 'bladder worms' (mixed in with black pudding) to a criminal on death row. Three days later the prisoner was duly executed, and Küchenmeister, on cutting into the man's guts, was delighted to find a crop of young tapeworms. Some years later he repeated this experiment, but this time fed the prisoner four months prior to his execution. This time the post-mortem dissection produced a tapeworm 150 cm (5 ft) long. If left undisturbed, tapeworms can reach lengths of 15 m (50 ft).

THE HUMANITARIAN ARGUMENT FOR POISON GAS

The British chemist Lyon Playfair proposed a method of breaking the stalemate at the Siege of Sevastopol during the Crimean War. His idea was to fire shells filled with cyanide against the Russian ships, but his suggestion was rejected by the British Ordnance Department, who regarded it as 'as bad a mode of warfare as poisoning the wells of the enemy'. In response, Playfair wrote:

> There was no sense in this objection. It is considered a legitimate mode of warfare to fill shells with molten metal which scatters among the enemy, and produces the most frightful modes of death. Why a poisonous vapour which would kill men without suffering is to be considered illegitimate warfare is incomprehensible. War is destruction, and the more destructive it can be made with the least suffering the sooner will be ended that barbarous method of protecting national rights. No doubt in time chemistry will be used to lessen the suffering of combatants, and even of criminals condemned to death.

THE FIRST RUBBER

The first rubber condoms were manufactured, 11 years after Charles Goodyear had patented his method of vulcanizing rubber. These rubber condoms were made by wrapping pieces of raw rubber round a penis-shaped mould, and then dipping the whole thing into a chemical solution. Following its invention in 1920, latex (a suspension of synthetic rubber in water) became the material of choice in condom manufacture, being stronger, thinner and longer-lasting than rubber. *See also* 1850 BC.

GOD PLAYS TRICKS

Philip Gosse, a noted biologist and fellow of the Royal Society, but also a member of the Plymouth Brethren and hence a Christian fundamentalist, published his book *Omphalos* (the Greek word for 'navel'), subtitled 'an Attempt to Untie the Geological Knot'. In this he sought to reconcile the chronology of the world provided in Genesis with the chronology suggested by the fossil and geological evidence, by proposing (in his son Edmund Gosse's words) 'that God hid fossils in the rocks in order to tempt geologists into infidelity'. He also affirmed that Adam, the prototype of all humans, had a navel. The purpose of this was the same as that of the fossils, Gosse argued: to give the appearance of an apparent past. 'We might still speak of the inconceivably long duration of the processes in question,' wrote Gosse, 'provided we understand ideal instead of actual time – that the duration was projected in the mind of God, and not really existent.' Of course, Gosse said, Adam's navel was never attached to an umbilical cord, as Adam had no mother. This account of a trickster God pleased neither the scientists nor the churches, and, as his son recalled, 'atheists and Christians alike looked at it and laughed, and threw it away'. Even Gosse Senior's friend Charles Kingsley, the clergyman and author of *The Water-Babies*, wrote 'I cannot . . . believe that God has written on the rocks one enormous and superfluous lie for all mankind.' Two years later Darwin published *On the Origin of Species*.

THE GREAT STINK

During the hot summer, London's inadequate drains failed to cope with the waste products of its population, resulting in what was known as the Great Stink. So distracting was the stench that MPs broke off their proceedings while the curtains of the House of Commons were soaked in chloride of lime to try to reduce the noxious effects of the effluvia emanating from the River Thames. The Great Stink did have one benefit, however. Parliament was persuaded that it was time to invest in a fully functional sewerage system for the capital.

A soot-covered chimneysweep's boy refuses the invitation of Father Thames to bathe in his noxious waters, filled as they are with slime and dead animals.

NOTHING TO REPORT

(May) Thomas Bell, president of the Linnaean Society, summed up the proceedings of the previous year thus:

> This year . . . has not, indeed, been marked by any of those striking discoveries which at once revolutionize, so to speak, the department of science in which they occur.

The significance of a joint paper, 'On the Tendency of Species to form Varieties; and on the Perpetuation of Varieties and Species by Natural Means of Selection', by Charles Darwin and Alfred Russel Wallace, read before the Linnaean Society on 1 July 1858, seems to have escaped Bell's notice.

THE SOLAR SUPERSTORM

The most powerful solar storm ever recorded took place between 28 August and 2 September, characterized by numerous sunspots and massive solar flares. The largest flare occurred just before midday on 1 September, and was observed by the British astronomer, Richard Carrington. The flare caused the ejection of a huge amount of plasma – ionized gas – which, travelling at some 8 million kph (5 million mph), reached the Earth some 18 hours later, causing a geomagnetic storm so powerful that it led to the failure of telegraph systems across Europe and North America. Auroras (which result from plasma being trapped by the Earth's magnetic field) were seen as far south as the Caribbean, and in places the lights were so bright that people could read outdoors at night. Others believed they were seeing the glow from burning cities. In the Rocky Mountains gold miners were woken up and started to prepare breakfast, thinking that it was sunrise – when it was only one o'clock in the morning. If a solar storm on such a scale were to occur today, the consequences would be considerably more serious, damaging satellites, knocking out radio communications and leading to power blackouts across the world. Records indicate that such events occur on average every 500 years.

LINOLEUM PREVENTS DISEASE

Linoleum was invented. One of the reasons for its subsequent popularity was that it was thought useful in preventing tuberculosis. It came to be believed that the germs responsible for TB lurked in the cracks between floorboards, so covering wooden floors with linoleum would, according to those advertising the merits of the new material, be a great health benefit.

TIRED OF THIS SORT OF THING CALLED SCIENCE

'I am tired of this sort of thing called science,' declared Senator Simon Cameron of Pennsylvania, opposing further funding for the Smithsonian Institution. 'We have spent millions on that sort of thing for the last few years, and it is time it should be stopped.'

THOSE MAGNIFICENT MEN IN THEIR FLYING MACHINES

The single-minded meteorologist James Glaisher, accompanied by the balloonist Henry Croxwell, ascended to a height of 11,270m (37,000ft or 7 miles) to make observations of temperature and humidity. They possibly had not intended to fly quite so high, as in the highly rarefied air Glaisher's vision began to blur, and then he passed out. Croxwell tried to release the gas valve with his hands, but they were so stiff with cold he could not, and so he was obliged to use his teeth. As they descended, he spoke the words 'temperature observation' into the ear of his unconscious colleague, who immediately came to, picked up his pencil and recommenced his recordings.

DON'T GIVE UP THE DAY JOB

Aleksandr Borodin, noted for his work on aldehydes, was appointed professor of chemistry at the Academy of Medicine in St Petersburg, and that year published a paper describing the first nucleophilic displacement of chlorine by fluorine in benzoyl chloride. Borodin is better known as the composer of such works as the opera *Prince Igor*, which provided the melody for the song 'I'm a Stranger in Paradise' in the musical *Kismet*. Other figures who have combined scientific and other vocations include:

- The French mathematician Pierre de Fermat (1601–65), who was also a criminal lawyer.
- The English physicist and mathematician Isaac Newton (1642–1727), who was also a member of Parliament, Warden of the Royal Mint and a speculative theologian.
- The German dramatist Georg Büchner (1813–37), author of *Danton's Death* and *Woyzeck*, who made a study of the barbel, a carp-like fish.
- Lewis Carroll (Charles Lutwidge Dodgson, 1832–98), the author of *Alice in Wonderland*, who was one of the leading British mathematicians of his time.
- The Russian novelist Vladimir Nabokov (1899–1977), who was an amateur lepidopterist specializing in American blues.
- The experimental American composer John Cage (1912–92), who was an expert in certain types of toadstool.

A large number of writers have been involved in the medical profession: Anton Chekhov, George Crabbe, Arthur Conan Doyle, Oliver Goldsmith, John Keats, François Rabelais, Tobias Smollett and William Carlos Williams, to name but a few.

PLOT TO ASSASSINATE LINCOLN WITH INFECTED LINEN

Dr Luke Blackburn, a supporter of the Confederacy during the American Civil War, sailed to Bermuda where at the time yellow fever was raging. There he collected the linen and clothes of those who had died of the disease, and arranged for them to be sent to the cities of the North, in the hope of initiating a devastating epidemic (not realizing that the disease is in fact spread by mosquito bites). It was said that some of the clothing was intended for President Lincoln himself. Blackburn remained in Canada to avoid prosecution, and later served as governor of Kentucky.

A PERPETUAL CLOCK?

Arthur Beverley constructed a clock that is still running today, even though it has never been wound up. The Beverley Clock is kept in the Department of Physics at Otago University, Dunedin, New Zealand, and its mechanism is driven by variations in atmospheric pressure and ambient temperature, which cause the air in an air-tight box to expand or contract, pushing on a diaphragm. A daily temperature variation of 6°C (11°F) is sufficient to raise a weight of 0.55 kg (1 lb) by 2.5 cm (1 in), which keeps the clock running.

GOD-GIVEN INCH VERSUS ATHEISTICAL FRENCH CENTIMETRE

Charles Piazzi Smyth, the Astronomer Royal for Scotland, conducted detailed measurements of the Great Pyramid at Giza. He concluded that it was designed using a unit he called the 'pyramid-inch', equal to 1.001 British inches and based on the Biblical cubit. He also asserted that the architect of the Great Pyramid must have been an Old Testament patriarch, perhaps the priest Melchizedek, who had been guided by God. The inch being therefore divinely ordained, Piazzi Smyth was confirmed in his fervent opposition to an official proposal made in 1864 to introduce the metric system into the UK – for was not the metre and all its kin the spawn of atheistical French radicals? In his researches, Piazzi Smyth found many 'significant' measurements in the structure of the Great Pyramid: for example, the perimeter of the base measured 365,000 inches, 1000 times the number of days in the year, while the height of the pyramid in inches bore a numerical relationship to the distance to the Sun measured in miles. Piazzi Smyth fervently believed that embedded in these and other measurements, numbers and ratios were a collection of prophecies, including a prediction that the world would come to an end in 1881. When this failed to occur, he issued a succession of alternative dates. The Royal Society declined to publish his paper on the matter, and this and other disappointments led to his resignation as Astronomer Royal for Scotland in 1888. Sadly, Piazzi Smyth's dabbling in pyramidology obscured his real scientific achievements in the fields of spectroscopy and infrared astronomy. He died in 1900, and is buried in the Yorkshire village of Sharow under a pyramid with a cross on its top. Piazzi Smyth was the uncle of Robert Baden-Powell, author of *Scouting for Boys*.

INSPIRED BY A DREAM

Friedrich August Kekulé von Stradonitz published a paper in which he outlined the structure of the benzene mol'ecule. This consists of a ring of six carbon atoms linked by alternating single and double bonds, and with a hydrogen atom attached to each atom of carbon. The formula of benzene had been known for some years, but its structure was a puzzle, and something of a holy grail for organic chemists. Then, one day during his time as professor of chemistry at the University of Ghent, Kekulé fell into a reverie:

> During my stay in Ghent, I lived in elegant bachelor quarters in the main thoroughfare. My study, however, faced a narrow side-alley and no daylight penetrated it . . . I was sitting writing on my textbook, but the work did not progress; my thoughts were elsewhere. I turned my chair to the fire and dozed. Again the atoms were gambolling before my eyes [he had had a similar experience some years before in London]. This time the smaller groups kept modestly in the background. My mental eye, rendered more acute by the repeated visions of the kind, could now distinguish larger structures of manifold conformation; long rows sometimes more closely fitted together all twining and twisting in snake-like motion. But look! What was that? One of the snakes had seized hold of its own tail, and the form whirled mockingly before my eyes. As if by a flash of lightning I awoke; and this time also I spent the rest of the night in working out the consequences of the hypothesis.

Kekulé later refined his hypothesis, suggesting that each bond between the carbon atoms was single half the time, and double half the time. He gave the above account of his dream in a speech to the German Chemical Society, who in 1890 arranged a celebration to mark the 25th anniversary of his groundbreaking paper.

In 1869, four years after Kekulé's dream, another major chemical breakthrough was inspired by a dream: this was Dmitri Mendeleev's discovery of the Periodic Table of the Elements.

DISPENSING WITH THE NECESSITY OF STUFFING

(27 May) Death of the eccentric naturalist Charles Waterton, who claimed to be descended from eight saints and the ancient Anglo-Saxon kings of England. His family had estates in what is now Guyana, where he embarked on a number of expeditions into the interior. These he recorded

'The Nondescript', a fanciful specimen based on the rear end of howler monkey, mischievously passed off by Charles Waterton as a wild man of the woods.

in a popular book, *Wanderings in South America*, which inspired the young Charles Darwin. To mount the specimens he brought back, Waterton developed a new technique of taxidermy in which he soaked the skins in a solution of alcohol and bichloride of mercury for some hours, then, after they had dried, he moulded them into the required shape from the interior, dispensing with the necessity of stuffing. He did not always display the results with full scientific seriousness: one specimen, a distorted rendition of a howler monkey, he labelled 'The Nondescript', while in a notable tableau he dressed a number of reptiles as famous Englishmen and entitled it 'The English Reformation Zoologically Demonstrated'. In 1817, on a visit to Rome (he was a devout Catholic), he climbed up the cross on the summit of the dome of St Peter's and left a glove on the lightning conductor on the top. The pope, Pius VII, asked him to remove it, which he duly did.

1865
Genetics

TOO GOOD TO BE TRUE?

The Austrian monk Gregor Mendel read his paper 'Experiments on Plant Hybridization' to the Natural History Society of Brünn, describing his work with breeding peas and outlining the laws of inheritance he thereby discovered. His paper – which laid the ground for the modern science of genetics – was published the following year in the *Proceedings* of the society, but was almost entirely forgotten until rediscovered in the early 20th century. Although Mendel's laws are firmly established, statisticians have expressed suspicion regarding his data. Where the laws stated the *probabilities* of the numbers in each generation possessing a certain characteristic, Mendel's results invariably recorded *exactly* these numbers. The odds against this happening are considerable. It seems that, one way or another, the Austrian monk doctored his data – a classic case of the phenomenon of confirmation bias, where people select information that confirms their preconceptions or hypotheses.

WHAT USE ELECTRICITY?

(25 August) Death of Michael Faraday, the discoverer of electromagnetic induction. When some years before he had been asked by W.E. Gladstone, then Chancellor of the Exchequer, what was the use of electricity, he had replied: 'Why sir, there is every possibility that you will soon be able to tax it.'

AN ADEQUATE DEGREE OF ACCURACY

At the age of just 25, Osborne Reynolds was appointed to the new chair of engineering at Owens College in Manchester. Reynolds was passionately interested in mechanics, especially fluid dynamics, and realized the importance of mathematics in attempting to understand phenomena in this field. But being of a practical bent, he concluded that in some contexts *absolute* accuracy was not necessary. When teaching an introductory class on the use of the slide rule, he took the simple example of 3 multiplied by 4. He took his class through the steps required to find an answer, and then announced: 'Now we arrive at the result; 3 times 4 is 11.8.' When the class sniggered, Reynolds put them right, declaring 'That is near enough for our purpose.'

OPERATING ON THE QUEEN'S ARMPIT

Joseph Lister, the pioneer of antisepsis, operated on Queen Victoria's armpit, where an abscess measuring 15 cm (6 in) had developed. He later claimed to be 'the only man who has ever stuck a knife into the Queen'.

A NEW AUSTRALIAN SPECIES?

1872 — Ichthyology

(August) When Carl Staiger, director of the Brisbane Museum, visited Gayndah Station in Queensland, he was served a hitherto unknown species of fish. Before eating it, he made a sketch and description, which he sent to the great French naturalist, François Louis de la Porte, Comte de Castelnau. Castelnau described the new species, named *Ompax spatuloides*, in 1879 in the *Proceedings of the Linnean Society of New South Wales*, and it continued to appear in lists of Australian fishes into the 1930s. It subsequently emerged that the 'fish' was a hoax, comprising the body of a mullet, the tail of an eel and the head of a needlefish.

A FOG AS DENSE AS IRON

1872 — Pathology

Louis Pasteur's germ theory met with extreme scepticism in some quarters. Professor Pierre Pochet, for example, declared:

> Pasteur's theory of germs is a ridiculous fiction. How do you think that these germs in the air can be numerous enough to develop into all these organic infusions? If that were true, they would be numerous enough to form a thick fog, as dense as iron.

THE 'BULLET PREGNANCY'

1874 — Reproductive science

LeGrand G. Capers, who had served as a physician in the Confederate army during the American Civil War, published in the *American Medical*

Weekly an account of a remarkable case of artificial insemination he said he had witnessed during the Battle of Raymond in 1863:

> Our men were fighting nobly, but pressed by superior numbers, had gradually fallen back to within one hundred and fifty yards of the house. My position being near my regiment, suddenly I beheld a noble, gallant young friend staggering closer, and then fall to the earth. In the same moment a piercing scream from the house reached my ear! I was soon by the side of the young man, and, upon examination, found a compound fracture, with extensive comminution of the left tibia; the ball having ricocheted from these parts, and, in its onward flight, passed through the scrotum, carrying away the left testicle. Scarcely had I finished dressing the wounds of this poor fellow, when the estimable matron came running to me in the greatest distress, begging me to go to one of her daughters, who, she informed me, had been badly wounded a few minutes before. Hastening to the house, I found that the eldest of the young ladies had indeed received a most serious wound. A minnie ball had penetrated the left abdominal parietes, about midway between the umbilicus and anterior spinal process of the ilium, and was lost in the abdominal cavity, leaving a ragged wound behind . . .
>
> About six months after her recovery, the movements of our army brought me again to the village of R., and I was again sent for to see the young lady. She appeared in excellent health and spirits, but her abdomen had become enormously enlarged, so much so as to resemble pregnancy at the seventh or eighth month . . . Just 278 days from the date of the receipt of the wound by the minnie ball, I delivered this same young lady of a fine boy, weighing eight pounds . . .
>
> About three weeks from the date of this remarkable birth, I was called to see the child, the grandmother insisting there was 'something wrong about the genitals'. Examination revealed an enlarged, swollen, sensitive scrotum, containing on the right side a hard, roughened substance, evidently foreign. I decided upon operating for its removal at once, and in so doing, extracted from the scrotum a minnie ball, mashed and battered as if it had met in its flight some hard, unyielding substance.

Prior to the birth, Dr Capers said he had examined the young lady, who had insisted she was a virgin, and indeed found her hymen was intact. She was subsequently introduced to the young man who had inadvertently fathered her child; the couple married, and proceeded to have two more children, this time in the orthodox fashion. Even as late as 1959, the *New York State Journal of Medicine* published a paper on 'Two Unusual Cases of Gunshot Wounds of the Uterus' that accepted the story as true. Capers had in fact intended the story as a satire on the many tall tales being told in the 1870s about the Civil War, and had submitted the article anonymously. However, E.S. Gaillard, the editor of the *American Medical Weekly*, had recognized Capers's handwriting, and published the account under his name.

FREEZING A MAN TO DEATH ON A WARM SUMMER'S DAY

(2 July) An article appeared in a Nevada newspaper, the *Territorial Enterprise*, describing how a Mr Jonathan Newhouse, 'a man of considerable inventive genius', had devised what he called 'solar armour' to protect him from the heat of the desert sun.

> The armour consisted of a long, close-fitting jacket made of common sponge and a cap or hood of the same material; both jacket and hood being about an inch in thickness. Before starting across a desert, this armour was to be saturated with water. Under the right arm was suspended an India rubber sack filled with water and having a small gutta-percha tube leading to the top of the hood. In order to keep the armour moist, all that was necessary to be done by the traveller, as he progressed over the burning sands, was to press the sack occasionally, when a small quantity of water would be forced up and thoroughly saturate the hood and the jacket below it. Thus, by the evaporation of the moisture in the armour, it was calculated might be produced almost any degree of cold.

Newhouse set off on foot to cross Death Valley in order to test his invention. The article described what happened next:

> The next day, an Indian who could speak but a few words of English came to the camp in a great state of excitement. He made the men understand that he wanted them to follow him. At the distance of about twenty miles out into the desert, the Indian pointed to a human figure seated against a rock. Approaching, they found it to be Newhouse still in his armour. He was dead and frozen stiff. His beard was covered with frost and – though the noonday sun poured down its fiercest rays – an icicle over a foot in length hung from his nose. There he had perished miserably, because his armour had worked but too well, and because it was laced up behind where he could not reach the fastenings.

The story was repeated in a number of other publications, including the *New York Times*, *Scientific American* and the *Daily Telegraph* of London. A more detailed update appeared in the *Territorial Enterprise* on 30 August, in which it was reported that among Newhouse's belongings, left in camp prior to his fatal foray, were:

> several bottles and small glass jars containing liquids and powders or salts of various kinds, with the nature of the most of which no person in the settlement was acquainted. One of the largest bottles was labelled 'Ether', known to them to be a very volatile liquid and capable of producing an intense degree of cold by evaporation.

Of course, the whole thing turned out to be a hoax, perpetrated by the newspaperman Dan DeQuille (pseudonym of William Wright), an associate and friend of Mark Twain. However, it was based on sound science, for over a century previously, in 1758, Benjamin Franklin and John Hadley (professor of chemistry at Cambridge University) had conducted experiments that showed that the evaporation of highly volatile liquids such as ether or alcohol could be used to bring the temperature of an object to below the freezing point of water. In their experiment they succeeded in this fashion to lower the temperature of a thermometer bulb to 7°F (−13°C) while the ambient temperature was 65°F (18°C). 'From this experiment,' Franklin wrote shortly afterwards to John Lining, 'one may see the possibility of freezing a man to death on a warm summer's day.'

DANGEROUS FORCES LET LOOSE

In the course of an inquiry into 'the so-called internal combustion engine', a US Congressman noted that 'The discovery with which we are dealing involves forces of a nature too dangerous to fit into any of our usual concepts.' Even by 1903, few had yet made the conceptual leap: 'Nothing has come along,' US businessman Chauncey Depew told his nephew, 'that can beat the horse and buggy.' He proceeded to advise his nephew against investing $5000 in the Ford Motor Company.

DOUBTS ON THE USEFULNESS OF THE TELEPHONE

When Alexander Graham Bell demonstrated his telephone to his father-in-law, Gardiner Greene Hubbard, the latter was not impressed. 'Hmph!' he expostulated, 'Only a toy.' President Rutherford B. Hayes was equally unimpressed: 'That's an amazing invention,' he told Bell, 'but who would ever want to use one of them?' In Britain, William (later Sir William) Preece, consulting engineer and later engineer-in-chief of the Post Office, remarked that 'Americans have need of this invention, but we do not. We have plenty of messenger boys.' President Hayes later had a telephone installed in his office, and hailed it as 'the greatest invention since Creation', while the mayor of Chicago predicted that before long there would be 'one in every city'. Some persisted in failing to see the point, however. When one US senator was told that the telephone would enable Maine to speak to Texas, he was distinctly underwhelmed. 'What,' he snorted, 'should Maine have to say to Texas?'

ON THE IMPORTANCE OF VERIFYING THAT DEATH HAS OCCURRED

A correspondent of the *British Medical Journal* in Naples reported a case that had come before the appeals court. It concerned a woman who was interred in the belief that she was dead, whereas she was only in a trance. Some days later, when the grave was opened for the reception of another body, 'it was found that the clothes which covered the unfortunate woman were torn to pieces, and that she had even broken her limbs in attempting to extricate herself from the living tomb'. The court sentenced the doctor who had certified the death and the major who had authorized the interment to three months' imprisonment for involuntary manslaughter.

DO HORSES FLY?

(11 June) Leland Stanford, governor of California, sponsored the photographer Eadweard Muybridge to settle the question as to whether a galloping horse is ever completely airborne, i.e. with all four hooves off the ground. Muybridge set up a series of 12 cameras evenly spaced along a distance of 6 m (20 ft) – the length of a horse's stride – alongside the track where the horse was to gallop. Each camera was triggered to take

a photograph at one thousandth of a second by a tripwire set off by the horse's hooves. The resulting photographs proved that a galloping horse is indeed briefly airborne, at the moment when all four hooves are tucked under the horse. Muybridge had been a bookseller until seriously injuring his head in a stagecoach accident in 1860, after which he suffered something of a personality change, becoming both more excitable and more inventive. His subsequent sequences of photographs of animals and humans in motion are important contributions to science – and to the development of cinematography.

Eadweard Muybridge's sequence of photographs – taken at intervals of 1/1000th second – proved once and for all that a galloping horse does lift all four hooves off the ground at the same time.

CONFUSING CORRELATION AND CAUSATION?

In *Nature*, the economist William Stanley Jevons published a paper entitled 'Commercial crises and sun-spots'. In this statistical study, he pointed to a correlation between changes in sunspot activity and terrestrial business cycles. He argued that sunspot activity had an effect on the Earth's weather, and that in turn affected crop yields. This in turn determined the price of food and other agricultural commodities such as cotton, and thus the economy as a whole. The link between sunspot activity and terrestrial weather has not been irrefutably established, although in 2003 two Israeli academics, astrophysicist Lev Pustilnik and agricultural economist Gregory Yom Din, correlated peaks in wheat prices in 17th-century England with periods of low sunspot activity. They went on to find similar correlations in US wheat prices in the 20th century. Other researchers have found correlations between sunspot activity, stock prices and the length of women's skirts – for example, Paul Macrae Montgomery in his 1975 paper, 'The Hemline Indicator of the Stock Market'.

INCANDESCENT BEARDS

Thomas Edison and his assistants at Menlo Park, New Jersey, the world's first industrial research laboratory, began the search for a material that would glow but not rapidly burn up when an electric current was passed through it. The quest took 14 months and several hundred experiments using all kinds of materials: cotton, jute, silk, manila hemp, horsehair, fishing line, teak, boxwood, spruce, vulcanized rubber, cork, celluloid, grass fibres, linen twine, tar paper, wrapping paper, cardboard, tissue paper, parchment, holly wood, rattan, California redwood and New Zealand flax, to name but a few.

Hairs from the beards of two of Edison's assistants were also tested, and bets placed on which would glow for the longest time before burning up. Those who bet on the loser complained that their man's hair had been subjected to a stronger current. Eventually, towards the end of 1879, Edison and his team came up with the ideal material: carbonized paper – paper baked in an oven until all that is left is the carbon framework. And thus was born the first practical electric light bulb.

DARWIN'S SOUL ABSORBED WITH WORMS

(23 November) Charles Darwin wrote to William Turner Thiselton-Dyer that 'My whole soul is absorbed with worms just at present.' The following year, just a few months before his death, he published *The Formation of Vegetable Mould through the Action of Worms with Observation of their Habits*, the result of more than 40 years of research. Unglamorous though worms might be compared to the explanatory power of evolution by natural selection, Darwin realized that worms play a key role in life on Earth by helping in the formation of soil. In the course of his researches, Darwin undertook some charmingly tangential experiments, concluding that 'Worms do not possess any sense of hearing.' He elaborated thus:

> They took not the least notice of the shrill notes from a metal whistle, which was repeatedly sounded near them; nor did they of the deepest and loudest tones of a bassoon. They were indifferent to shouts, if care was taken that the breath did not strike them. When placed on a table close to the keys of a piano, which was played as loudly as possible, they remained perfectly quiet. Although they are indifferent to undulations in the air audible by us, they are extremely sensitive to vibrations in any solid object. When the pots containing two worms which had remained quite indifferent to the sound of the piano, were placed on this instrument, and the note C in the bass

clef was struck, both instantly retreated into their burrows. After a time they emerged, and when G above the line in the treble clef was struck they again retreated. Under similar circumstances on another night one worm dashed into its burrow on a very high note being struck only once, and the other worm when C in the treble clef was struck.

BOILED ELEPHANT TRUNK AND STEWED MOLE

(19 December) Death of the naturalist Frank Buckland. The son of William Buckland, the distinguished geologist and canon of Christ Church, Oxford, young Frank was introduced by his father to unusual dishes, such as ostrich, horse tongue, squirrel pie, and mice in batter. While a student at Christ Church in the 1840s, he kept a number of exotic pets, including an eagle, a marmot, a monkey and a bear, which was introduced on one occasion to the geologist Charles Lyell. It was while at Oxford that Buckland began his own career as an experimental zoophage. Having heard of the death and burial of the panther at Surrey Zoological Gardens, he obtained the corpse and cooked and ate it. He later recalled that, having been buried for two days, 'It was not very good.' In 1860 he set up the Acclimatization Society, to investigate which animals from around the world might be reared in Britain for the table. At their dinners the members of the society tried out a wide range of dishes, including kangaroo, sea slug, rhinoceros ('like very tough beef'), boiled elephant trunk ('rubbery'), porpoise head ('broiled lamp-wick'), giraffe (like veal, apparently; the unfortunate animal had been roasted when its quarters at London Zoo caught fire) and stewed mole ('utterly horrible').

Buckland was also a keen pisciculturist, and in 1867 was appointed Inspector of Salmon Fisheries. His interest in fish ranged far beyond salmon, and when he heard that a fishmonger in Bond Street had obtained a 2.75 m (9 ft) sturgeon, he begged to borrow it for the night to make a cast. He struggled to get it home, and then began to lower it down the basement steps with a rope tied round its tail. But the fish was one that got away, as he described in an article in *Land and Water* (27 April 1867):

> He started all right, but 'getting way' on him, I could hold the rope no more, and away he went sliding headlong down the stairs, like an avalanche down Mont Blanc . . . he smashed the door open . . . and slid right into the kitchen . . . till at last he brought himself to an anchor under the kitchen table. This sudden and unexpected appearance of the armour-clad sea monster, bursting open the door . . . instantly created a sensation. The cook screamed, the house-maid fainted, the cat jumped on the dresser, the dog retreated behind the copper and barked, the monkeys went mad with fright, and the sedate parrot has never spoken a word since.

1882
Physics

A LATE DEVELOPER

Albert Einstein, aged three and a half, uttered his first words, to complain that his milk was too hot. His parents, who thought their child was severely retarded, were delighted. When they asked him why he hadn't spoken before, he told them that there had been no need, as hitherto everything had been in order.

1883
Geology

A BIGGER BANG

(27 August) The volcanic eruption of Krakatoa in Indonesia almost destroyed the entire island in a series of four massive explosions. The last and biggest, at 10.02 a.m., is estimated to have been the equivalent of a 150-megaton nuclear explosion (the biggest hydrogen bomb ever tested had a yield of

50 megatons), and could be heard 4600 km (2850 miles) away, on the Indian Ocean island of Rodriguez. The pressure wave was recorded on barographs around the Earth, bouncing backward and forward seven times. Ash from the eruption was propelled to a height of 80,000 m (50 miles), causing total darkness in the region for more than two days, and landed over an area of 800,000 square km (300,000 square miles). Some 21 cubic km (5 cubic miles) of rock was thrown into the air, although most of the island actually sank beneath the waves. The seas round about were rendered unnavigable for a while by thick swathes of floating pumice. Krakatoa was uninhabited, and few people died in the actual eruption, but the tsunamis generated by the explosions initially had a height of 40 m (130 ft), and killed 40,000 people along the coasts of Java and Sumatra. The waves were still 1 cm (0.4 in) high when they arrived at Le Havre in France, 32 hours later.

1884
Medicine

A CASE OF PENIS CAPTIVUS

The 13 December issue of the *Philadelphia Medical News* carried a letter from a Dr Egerton Yorrick Davis, a retired surgeon captain of the US Army, reporting 'an uncommon form of vaginismus' (sudden tensing of the vaginal muscles). This had been suffered during intercourse by a maid Dr Davis had been called upon to attend, with the consequence that the penis of her lover, the coachman, was trapped inside her. Luckily, the administration of chloroform to the embarrassed maid had relaxed her sufficiently for the coachman to withdraw. The case was, in fact, a fiction, created by Sir William Osler, pioneer of modern medical education at Johns Hopkins University and one of the most distinguished physicians of his age (he went on to hold the Regius Chair of Medicine at Oxford). Osler had a puckish sense of humour, and, under the Davis pseudonym, contributed a number of reports of fantastical medical cases – usually relating to sexual pathology – to various journals.

SUPERNATURAL INSTRUCTION ON THE VALUE OF PI

An amateur mathematician, Edward Johnston Goodwin, who practised medicine in Solitude, Indiana, claimed that the value of pi was exactly 3.2. 'During the first week of March, 1888,' he later recalled, 'the author was supernaturally taught the exact measure of the circle . . . no authority in the science of numbers can tell how the ratio was discovered.' His paper, 'Universal Inequality is the Law of All Creation', was published under the title 'Quadrature of the Circle' in 1894 in *American Mathematical Monthly*, 'at the request of the author'. In 1897 Goodwin's 'proof' was enshrined in 'A bill for an act introducing a new mathematical truth and offered as a contribution to education to be used only by the state of Indiana free of cost'. This bill was introduced, at Goodwin's urging, into the Indiana House of Representatives. The House at first referred the bill to the Committee on Swamp Lands, who in turn referred it to the Committee on Education, who recommended that 'said bill do pass'. The House duly complied, and the bill was forwarded to the State Senate, where it was referred to the Committee on Temperance, who also recommended that 'said bill do pass'. However, as the *Indianapolis Journal* reported, the Senators were not entirely convinced:

> Senator Hubbell characterized the bill as utter folly. The Senate might as well try to legislate water to run up hill . . . All of the Senators who spoke on the bill admitted that they were ignorant of the merits of the proposition. It was simply regarded as not being a subject for legislation.

As a consequence, consideration of the bill was indefinitely postponed, and so never became law.

BALLYBUNION POINTS THE WAY TO THE FUTURE

In Ireland one of the world's first monorail systems was opened, running between Listowel and Ballybunion in County Kerry. The railway was designed by the French engineer Charles Lartigue, who while in North Africa had been inspired by the sight of camels carrying panniers on each side of their humps. The Listowel–Ballybunion line accordingly had a single load-bearing rail in the centre, with two lower rails at either side along which ran stabilizing wheels. Balance was indeed a key consideration: if a farmer wanted to send a cow to market, he would be obliged to send two calves along with it, placed on the other side of the wagon to the cow; on the return journey, the two calves would be carried back, one on each side. The monorail remained in service until 1924, when it was closed following damage incurred in the Civil War. A short section has been restored.

THE BROWN–SÉQUARD ELIXIR

In his old age, the Mauritian-born physician Charles-Édouard Brown-Séquard, who was already celebrated for his discoveries in neurology and physiology, reported that injecting himself with extracts taken from the testicles of guinea pigs and dogs put a spring in his step. Now, he claimed, he could run up stairs as in his youth, and, what was more, the stream of his urine, when he passed water, extended beyond its previous length by some 25 per cent. 'I should add,' he coyly added, in an address to the Biological Society of Paris on 1 June, 'that other powers, which I admit had not deserted me completely but which were without doubt not what they once were, have also shown a significant increase in vigour.' He thus proposed the procedure as a means of prolonging youth, vitality and life itself – but other scientists were sceptical of what they dubbed the 'Brown-Séquard Elixir', and his reputation never recovered.

CANNABIS A PALLIATIVE FOR MORAL SHOCK

(4 July) The *British Medical Journal* reported that doctors frequently prescribed *Cannabis indica* (Indian hemp) for 'a form of insanity peculiar to women, caused by mental worry or moral shock, in which it clearly acts as a psychic anodyne'. The hemp, the *BMJ* reported, 'seems to remove the mental distress and unrest'. In 1902, the *Compendium of Materia Medica, Therapeutics and Prescription Writing* was recommending Indian hemp 'for a variety of colourful conditions', particularly those associated with the uterus, as the drug's 'powers as an anodyne and stimulant of the uterine muscular fibre render it a very efficient agent'.

THE CONSEQUENCES OF LIBERTINE BEHAVIOUR

In *The Laws and Mysteries of Love*, Alexandre Weill, a writer and polemicist much admired by Victor Hugo, gave his readers this stern warning: 'Every man who has sexual relations with two women at the same time risks syphilis, even if the two women are faithful to him, for all libertine behaviour spontaneously ignites this disease.'

ZEBRAS USED TO DRAW CARRIAGE

Walter Rothschild, scion of the wealthy banking family, opened his Natural History Museum at Tring, in Buckinghamshire, to the public. Here he trained zebras – hitherto regarded as intractable – to draw his carriage, and on one occasion rode thus to Buckingham Palace. Rothschild also succeeded in crossing zebras with horses, to produce 'zebroids'. His niece, Dame Miriam Rothschild, became a distinguished entomologist, and delighted in explaining to princes and statesmen that the colourful Christmas card she had sent them was not a detail of an impressionist painting, but the greatly magnified reproductive organ of a butterfly.

QUEEN VICTORIA RUMOURED TO TAKE CHINESE MEDICINE

European scientists studying an outbreak of plague in Hong Kong were subject to attacks by the native population, who believed they were going to be ground into powder as a medicine for the British royal family.

X–RAY–PROOF UNDERWEAR

Wilhelm Röntgen discovered X-rays, which were promptly denounced by Lord Kelvin as a hoax. Even Röntgen himself, devoted as he was to classical physics, was not that happy with the implications of his discovery. Soon, however, confirmation of the existence of X-rays came from laboratories

A FALSTAFFIAN TREE IN THE HAYMARKET,
AS SEEN BY RÖNTGEN RAYS

An 1896 cartoon from Punch, *showing how X-rays could prove the old adage that inside every fat man a thin man is trying to get out.*

all around the world. The idea of being able to see people's skeletons and inner organs was disturbing to many, and a newspaper in Graz, Austria, reported that a Professor Czermak had not been able to sleep since seeing his own 'death's-head'. In Paris, meanwhile, a Dr Baraduc claimed to have used X-rays to photograph the human soul. Others found the possibility of seeing 'through' solid material – such as clothing – inspired their prurient interest, hence the following rhyme:

I'm full of daze
Shock and Amaze;
For now-a-days
I hear they'll gaze
Thru' cloak and gown – and even stays,
Those naughty, naughty Roentgen Rays.

Subsequently, an enterprising firm in London advertised 'X-ray proof underclothing for ladies'.

A LIFT INTO SPACE

The Russian rocket pioneer Konstantin Tsiolkovsky visited Paris, and was inspired by the newly constructed Eiffel Tower to suggest an inexpensive means of space travel: a tower so high that it would reach into space. Within the tower, Tsiolkovsky proposed, would be an elevator that would lift passengers into geostationary orbit at an altitude of 35,790 km (22,190 miles) above the Earth. Tsiolkovsky was a disciple of the mystic Nikolai Fyodorov, who believed that colonizing other planets would bring about the perfection, and immortality, of the human race. In the century and more since, there have been a number of other proposals for space elevators, including some recent ones involving carbon nanotube technology and backed by NASA. Work continues among a number of groups to overcome various engineering limitations, and in 2008 a conglomerate of Japanese companies and universities announced that a space elevator could be made a reality for as little as $7.5 billion.

STUFF AND NONSENSE – PART I

Lord Kelvin, the grand old man of Victorian physics, pronounced that 'Heavier-than-air flying machines are impossible. I have not the smallest molecule of faith in aerial navigation other than ballooning.' The first manned flight in a powered heavier-than-air machine took place at Kitty Hawk just seven years later, on 17 December 1903, when the Wright brothers' *Flyer I* made three short flights, about 3 m (10 ft) above the ground.

A CHILD PRODIGY

At the age of three, J.B.S. Haldane – later celebrated as a geneticist and evolutionary biologist – is said to have looked at the blood oozing from a cut in his forehead and asked, 'Is it oxyhaemoglobin or carboxyhaemoglobin?'

NO FUTURE IN RADIO

Lord Kelvin saw little potential in Marconi's experiments in wireless transmission. 'Radio has no future,' he said.

A NEW COUGH MIXTURE

The German pharmaceutical company Bayer began to market a new synthetic drug, diacetylmorphine, which they claimed did not have the

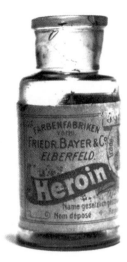

In 1898 the German pharmaceutical company Bayer launched a new over-the-counter cough mixture, which was held to be entirely non-addictive.

addictive properties of morphine itself. Diacetylmorphine was primarily promoted as a non-prescription cough suppressant, and also as an effective cure for morphine addiction. The brand name under which it was marketed derived from the German word *heroisch* ('heroic'): Heroin.

Two years after its launch, in the *Boston Medical and Surgical Journal*, James R.L. Daly was giving the new drug a warm endorsement. 'It possesses many advantages over morphine,' he wrote. 'It is not hypnotic and there is no danger of acquiring the habit.'

TOBACCO IS GOOD FOR YOU

In *The Doctor at Home*, George Black wrote that 'The power of tobacco to sustain the system, to keep up nutrition, to maintain and increase the weight, to brace against severe exertion, and to replace ordinary food, is a matter of daily and hourly demonstration.'

COCAINE AND TESTICLES

The German surgeon August Bier used his assistant, Dr Hildebrandt, as a guinea pig for his new technique of spinal anaesthesia, which involved injecting cocaine directly into the spinal fluid. Seven minutes after injecting him, Bier pricked Hildebrandt in the thigh with a needle. The latter felt this only as pressure. He then jabbed him in the thigh with a scalpel. Hildebrandt felt nothing. After 13 minutes, Bier put the burning tip of a cigar to Hildebrandt's thigh, eliciting no complaint. The numbing effect only worked below the waist: 'Pulling out pubic hairs was felt in the form of elevation of a skinfold; pulling of chest hair above the nipples caused vivid pain.' Elated, Bier set about Hildebrandt's shins with a hammer, squeezed his testicles and then stabbed his thigh right through to the bone. Hildebrandt felt nothing. Once the anaesthetic wore off, after about 45 minutes, the two went out to dinner to celebrate, drinking quantities of wine and smoking several cigars. Bier later became known for his dictum: 'Medical scientists are nice people, but you should not let them treat you.'

PITCHBLENDE AND GOOSEBERRY JELLY

In the same year that she discovered a new radioactive element, polonium, in a particular fraction of the mineral pitchblende, Marie Curie also noted down her recipe for gooseberry jelly:

> I took four kilos of fruit and the same weight in crystallized sugar. After boiling for ten minutes, I passed the mixture through a rather fine sieve. I obtained 14 pots of very good jelly, not transparent, which 'took' perfectly.

It is clear from this that she took as much care with her culinary chemistry as she did with her principal scientific work. By the end of the year she and her husband Pierre announced that they believed that pitchblende contained another new element: radium. It was to take another four years, however, before Marie was able to isolate a sample. It was not just a matter of heating little test tubes above a Bunsen burner. Pure pitchblende was very expensive, but the Austrian government had gladly donated a tonne of spoil from the slagheap of a pitchblende mine in Bohemia to these 'French lunatics', and Marie was to spend years in the courtyard of their home 'stirring a boiling mass, with an iron rod nearly as big as myself'. By each evening, she recalled, 'I was broken with fatigue.' At last, in 1902, she succeeded in preparing 0.1 g ($^{1}/_{280}$ oz) of pure radium. The fatal damage that the radioactivity was doing to her health did not become apparent until much later.

1900 to 1924

Nothing new to be discovered * The
poison squad * The rays that weren't *
Nuclear apocalypse foretold * Kissing
perverted, suggests Freud * Soul
weighs 21 grams * Communicating
with Mars * A stench odious to humans
but pleasing to cows * An end to plum
pudding * Return of Halley's Comet
augurs end of life as we know it * The
worst journey in the world * The long
road to Ecstasy * Testicle transplants
* Bohr's bosh * Einstein's excesses * A
new use for conkers * Madness, murder
and mathematics * 'A cure for the living
dead' * Edison's spirit communicator *
Genius or ignorant numbskull?

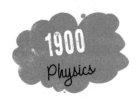

NOTHING NEW
TO BE DISCOVERED

Lord Kelvin pronounced the end of physics: 'There is nothing new to be discovered in physics now,' he declared. 'All that remains is more and more precise measurement.'

THE POISON SQUAD

Harvey W. Wiley, chief chemist in the US Department of Agriculture, embarked on a five-year study to test the effects of various chemical preservatives added to foodstuffs on a dozen healthy young male volunteers. This so-called 'Poison Squad', whose members were replaced at regular intervals, was fed steadily increasing doses of such substances as borax, formaldehyde, sodium benzoate and sulphuric acid, resulting in what the *New York Times* (22 May 1904) called 'pronounced inflammation of the digestive tract', accompanied by nausea, vomiting, stomach cramps and extreme lassitude. The *Times* went on to describe the procedure employed by Professor Wiley:

> At first poison was put into the food, without the squad knowing which dish was doctored. The effect upon the minds of the squad, whose stomachs finally began to rebel at the best of food, even that unpoisoned, caused a change in the method, and the poison was administered in capsules. The amount was increased and decreased as the experimenters deemed the system of the individual could stand it.
>
> All through the experiment the men were watched. They were examined three times a day. When health began to break there was a brief release from the poison diet. Today the men are thinner than usual, and all show the effects of the strain.

'What we tried to learn and did learn,' said an official of the department today, 'was the effect of food preservatives upon the system. This effect was mildly injurious or deadly, according to the amount and character of the preservatives absorbed . . .'

The public were both fascinated and appalled by the experiments, and the heroic young men were celebrated by Lew Dockstader's Minstrels in 'The Song of the Poison Squad', the chorus of which goes:

O, they may get over it but they'll never look the same,
That kind of bill of fare would drive most men insane.
Next week he'll give them mothballs, *à la* Newburgh or else plain;
O, they may get over it but they'll never look the same.

The publicity attending the study led to the successful passage of the Pure Food and Drugs Act of 1906, and the establishment of the Food and Drug Administration. The guinea pigs appear to have suffered no long-term harm: one of them died in 1979 at the grand old age of 94.

THE RAYS THAT WEREN'T

The French physicist René-Prosper Blondlot, while attempting to polarize the recently discovered X-rays, detected a new kind of radiation, which he dubbed 'N-rays', after the University of Nancy, where he worked. Scores of other scientists went on to detect the rays, which appeared to be emitted by many substances, including the human body. However, some of the leading physicists of the day, including Lord Kelvin, failed to repeat the experiments, and remained sceptical. The journal *Nature* dispatched the American physicist Robert Wood – who had also failed to detect N-rays – to Nancy to investigate. Wood was something of a practical joker – on a honeymoon visit to Yellowstone in 1892 he had secretly poured some fluorescent dye into a geyser, and watched as a few minutes later a group of tourists stood open-mouthed when the geyser erupted in a jet of brilliant green water and steam. In Blondlot's laboratory, Wood surreptitiously removed a crucial prism from the apparatus, and replaced a piece of material said to emit N-rays with a piece of wood, but still the French experimenters said they detected N-rays. *Nature* then published Wood's findings, which concluded that the detection of the N-rays was entirely dependent on the expectations and subjective perceptions of the experimenters. The desire of the French – defeated in the Franco-Prussian War of 1870 and smarting after the recent discovery of X-rays by the German physicist Wilhelm Röntgen – to reclaim the honour of France may also have played a significant part in the 'discovery' of the non-existent rays.

1903
Mathematics

NOT A PRIME AFTER ALL

The French monk, mathematician, musical theorist and theologian Martin Mersenne (1588–1648) had famously proposed that $2^{67} - 1$ was one of the so-called 'Mersenne primes', which are prime numbers that take the form $2^n - 1$, where *n* itself is prime. But not all such numbers *are* prime. At the meeting in New York of the American Mathematical Society in October 1903, F.N. Cole was due to read a paper entitled 'On the Factorization of Large Numbers'. When it came to his turn to speak, the notoriously taciturn Cole approached the blackboard and wrote out the arithmetic for raising 2 to its 67th power. Then he subtracted 1. Still in silence, he then moved to a clear part of the board and proceeded to carry out a laborious piece of long multiplication:

$$193,707,721 \times 761,838,257,287$$

The result of the two calculations was the same. The audience broke out into spontaneous applause – an unprecedented occurrence at a meeting of the American Mathematical Society. Cole resumed his seat, still without saying a word. There were no questions. Some time afterwards, when Cole was asked how long it had taken him to find the factors, he tersely replied, 'Three years of Sundays.' Mersenne himself had said that all eternity would not be enough to determine if a number of 15 or 20 digits was prime.

Discoveries regarding Mersenne primes seem to inspire celebration: when Donald B. Gillies of the University of Illinois found the 23rd Mersenne prime in 1963 using the ILLIAC II supercomputer, the university franked its letters with a special postmark proudly declaring '$2^{11213} - 1$ is prime'.

1904

Physics

NUCLEAR APOCALYPSE FORETOLD

Following his work with Ernest Rutherford on radioactive decay, Frederick Soddy gave a lecture on the dangers of this new source of energy: 'The man who put his hand on the lever by which a parsimonious Nature regulates so jealously the output of this store of energy would possess a weapon by which he could destroy the Earth if he chose.' Just ten years later H.G. Wells published *The World Set Free* (1914), in which biplanes dropped atomic bombs in some future war – a concept directly inspired by the work of Soddy, Rutherford and Sir William Ramsay, as Wells acknowledged:

> The problem which was already being mooted by such scientific men as Ramsay, Rutherford, and Soddy, in the very beginning of the twentieth century, the problem of inducing radio-activity in the heavier elements and so tapping the internal energy of atoms, was solved by a wonderful combination of induction, intuition, and luck by Holsten so soon as the year 1933.

The physicist Leó Szilárd read Wells's book in 1932, and the following year came up with the idea of the chain reaction – the key to nuclear power, and the atomic bomb (*see* 1933).

1905
Psychology

KISSING PERVERTED, SUGGESTS FREUD

Sigmund Freud published *Three Essays on the Theory of Sexuality*, in which he noted that certain preliminaries to copulation that humans linger over and take pleasure in may, on the grounds that they involve parts of the body apart from the reproductive organs, be considered as 'perversions':

> . . . the kiss, one particular contact of this kind, between the mucous membrane of the lips of the two people concerned, is held in high sexual esteem among many nations (including the most highly civilized ones), in spite of the fact that the parts of the body involved do not form part of the sexual apparatus but constitute the entrance to the digestive tract.

He goes on to distinguish between what may technically be defined as a 'perversion', and what by convention or by temperament may elicit 'disgust'. For example, 'a man who will kiss a pretty girl's lips passionately, may perhaps be disgusted at the idea of using her toothbrush, although there are no grounds for supposing that his own oral cavity, for which he feels no disgust, is any cleaner than the girl's'. Furthermore, those who are disgusted at the thought of sexual contact involving the anus on the grounds that that orifice comes into contact with excrement 'are not much more to the point than hysterical girls who account for their disgust at the male genital by saying that it serves to void urine'. Freud concludes: 'No healthy person, it appears, can fail to make some addition that might be called perverse to the normal sexual aim; and the universality of this finding is in itself enough to show how inappropriate it is to use the word perversion as a term of reproach.'

SOUL WEIGHS 21 GRAMS

(March) Both the *New York Times* and the *Washington Post* reported on some remarkable experiments that had been carried out over the previous few years by Dr Duncan MacDougall of Haverhill, Massachusetts. Anxious to prove the physical existence of the soul, Dr MacDougall devised an experiment to weigh it. At the Cullis Free Home for Consumptives in Dorchester, Massachusetts, he found a number of male volunteers who were dying of tuberculosis. As their last hours approached, MacDougall placed his subjects, still in their beds, on an enormous beam balance to see whether there was any movement at the moment of death. When his first subject breathed his last, MacDougall reported, 'suddenly coincident with death the beam end dropped with an audible stroke hitting against the lower limiting bar and remaining there with no rebound'. To rebalance the scales, MacDougall had to place two dollar pieces together weighing □ oz (21 grams) on the end with the late patient. A further five such experiments proved inconclusive, however. Nevertheless, MacDougall was delighted to note that when he tried the experiment 15 times on dogs, in no case was any loss of weight noted at death, proving to MacDougall's satisfaction that animals are not possessed of souls.

COMMUNICATING WITH MARS

The American physicist Robert Wood – who had debunked N-rays (*see* 1903) – constructed the first practical liquid-mirror telescope, using a rotating bowl of mercury as a parabolic reflector. This was placed down a dry well, and the images of the stars overhead were projected into the air about a metre (3 ft) above the mouth of the pit. So impressed were a group of amateur astronomers from Texas that they offered Wood $50,000 to construct a set of giant mercury mirrors to reflect sunlight and project signals towards Mars – which, on the basis of the 'canals' observed on its surface, many believed to be inhabited by a highly advanced civilization. Wood himself was sceptical, and somewhat mischievously suggested an alternative technique to attract the attention of the Martians. Why not, he proposed, create a huge black spot on the surface of the Earth made up of strips of black cloth. Every now and again, the strips could be rolled up, then unrolled again, so that the spot would appear to wink in the direction of the distant red planet.

A STENCH ODIOUS TO HUMANS BUT PLEASING TO COWS

At Cambridge University Sir William Jackson Pope was conducting experiments involving the non-metallic element selenium. One of the products of these experiments was hydrogen selenide – a compound with a smell even more unpleasant than the rotten-egg odour of its close relative, hydrogen sulphide. Such was the unpleasantness of the operation – for hydrogen selenide not only smells of decayed horseradish, but is also highly irritating – that Pope and his team abandoned the close atmosphere of the laboratory for the fresh air of the roof. It was June, and around Cambridge people were enjoying tea in their gardens. But not for long. Soon the air became laden with an invisible malevolence, and the genteel enjoyment of cucumber sandwiches was utterly spoilt. Across the city there was indignation and uproar, prompting the *Cambridge Daily News* to get to the bottom of it. 'SCIENCE THE SINNER,' its headline screamed, pointing the finger at the University Chemical Laboratory. Thereafter, the experiments were conducted in the depths of the fens, with the permission of the local farmer. One of Pope's team, John Read, later professor of chemistry at

St Andrews, recalled that the farmer soon fled, but 'A large herd of cows formed into a semicircle to leeward and provided a silent but appreciative audience . . . insects of many kinds swarmed over the apparatus, some of them even making determined attempts to force a way past the stoppers into the flasks. Their whole behaviour indicated that they felt they were missing something really good.'

AN END TO PLUM PUDDING

At the University of Manchester, Hans Geiger and Ernest Marsden, under the direction of Ernest Rutherford, conducted an experiment to investigate the structure of the atom. In 1904 J.J. Thomson, who had discovered the electron in 1897, concluded that the atom was like a plum pudding, with the negatively charged electrons (or 'corpuscles' as he called them) embedded like plums in a 'pudding' of positive charge. In the Geiger-Marsden experiment, a beam of alpha particles was fired at a thin sheet of gold foil. According to the plum-pudding model, the alpha particles should have passed through the foil undisturbed. However, Geiger and Marsden recorded that a small number of alpha particles were deflected. Rutherford was astonished:

> It was quite the most incredible event that has ever happened to me in my life. It was almost as incredible as if you fired a 15-inch shell at a piece of tissue paper and it came back and hit you.

Rutherford could only conclude that 'the greater part of the mass of the atom was concentrated in a minute nucleus. It was then that I had the idea of an atom with a minute massive centre, carrying a charge.' Thus Rutherford conceived the 'planetary' model of the atom: in this model most of an atom is empty space, in which the electrons orbit the nucleus. In terms of relative dimensions, compared to the whole atom, the nucleus is like a pea in a cathedral, although it contains more than 99.9 per cent of the atom's mass. This model in turn was superseded by the less easily graspable quantum model, in which the electrons form more of a cloud around the nucleus.

RETURN OF HALLEY'S COMET AUGURS END OF LIFE AS WE KNOW IT

(February) Observing the approach of Halley's Comet, astronomers at the Yerkes Observatory in Chicago announced that they had detected cyanogen, a colourless toxic gas with a pungent odour, in the comet's tail. This led the French astronomer and popularizer of science Camille Flammarion to predict that the gas 'would impregnate the atmosphere and possibly snuff out all life on the planet'. Despite reassurances by other astronomers such as Percival Lowell that the gases in the comet's tail were 'so rarified as to be thinner than any vacuum', there was widespread panic and a boom in sales of gas masks and so-called 'comet pills'. In parts of the United States, people prepared sealed rooms, even blocking keyholes with paper, while one man asked his friends to lower him down a deep, dry well, accompanied by a gallon of whiskey. The *Chicago Tribune* marked the uneventful passing of the comet in May with the wry headline 'We're Still Here.'

THE WORST JOURNEY IN THE WORLD

A three-man team comprising Edward Wilson, Apsley Cherry-Garrard and 'Birdie' Bowers made a gruelling 19-day round trip of 214 km (124 miles) through the darkness and blizzards of the Antarctic winter from Cape Evans to Cape Crozier and back. In temperatures as low as -60°C (-77°F), the men shivered so violently at night that they feared their bones might break. Cherry-Garrard, in his celebrated account, *The Worst Journey in the World*, said 'anybody would be a fool who went again'. The purpose of the journey was scientific – to retrieve some specimen Emperor penguin eggs, as Wilson

Edward Wilson, Apsley Cherry-Garrard and 'Birdie' Bowers prior to their terrible journey through the Antarctic winter in pursuit of Emperor penguin eggs.

believed that embryonic evidence might throw light on his hunch that the flightless penguins were a kind of missing link between reptiles and birds. The men were astonished to find that it was the males who incubated the eggs – and some eggless males were so desperate to fulfill their role that they fashioned lumps of ice into rough egg shapes, and sat on those. Having gathered three eggs, the men set off back to Cape Evans, but were only just alive when they struggled back to base camp. As part of the same expedition, Cherry-

Garrard played a supporting role in Captain Scott's push to the South Pole. His companions, Wilson and Bowers, went all the way with Scott, and after they failed to return, Cherry-Garrard joined the rescue effort that found the frozen corpses of the three men, huddled together in their tent.

When Cherry-Garrard eventually turned up with his three precious eggs at the Natural History Museum back in London, he was asked, 'Who are you? What do you want? This ain't an egg shop.' He was then left waiting for hours until someone could be bothered to give him a receipt. As it turned out, biology had moved on. Wilson had based his hunch on Ernst Haeckel's 1866 doctrine, 'Ontogeny recapitulates phylogeny', i.e. the embryonic development of an individual recapitulates the evolutionary development of its species. This had become discredited, and Cherry-Garrard was additionally told that the eggs added little to the museum scientists' knowledge of penguin embryology.

PILTDOWN MAN

(18 December) Charles Dawson, an amateur archaeologist, announced at a meeting of the Geological Society of London that four years earlier he had been presented by a workman with some ancient skull fragments found in a gravel pit at the Sussex village of Piltdown. He had visited the site himself and found further fragments, which he showed to Arthur Smith Woodward, keeper of the geological department of the British Museum. Woodward told the meeting that he and Dawson had found more items, and that he had made a reconstruction of the skull and lower jaw. His conclusion was that *Eoanthropus dawsonii* – 'Dawson's dawn-man' – represented the much sought-after 'missing link', providing evidence of Darwin's belief that humans had evolved from ape-like ancestors.

Almost immediately, there were suspicions among some scientists that 'Piltdown Man' (as it was popularly known) was a fake, and in 1923 the German anatomist and physical anthropologist Franz Weidenreich studied the remains and concluded that they comprised a modern human cranium and the jaw of an orang-utan, with the teeth filed down. But for the most part Britain's scientific establishment – perhaps proud that this early human ancestor was an Englishman – appears to have suspended its critical faculties, and it was not until 1953 that Piltdown Man was finally and definitively shown to be a hoax. The remains, it turned out, comprised a human skull of medieval date, the lower jaw of an orang-utan and some filed chimpanzee teeth, all stained with potassium dichromate to give the appearance of great age.

Various perpetrators of the hoax have been suggested. Dawson himself (who died in 1916) is a leading candidate – his own antiquarian collection was shown to contain dozens of fakes. But it is possible Dawson was just very gullible – and perhaps the dupe of M.A.C. Hinton, keeper of zoology at the British Museum, and an arch-rival of Smith Woodward. The evidence pointing towards Hinton – a well-known practical joker – takes the form of an old trunk, found in 1996 by workmen under the roof of the Natural History Museum. The trunk belonged to Hinton, and was found, when examined by Brian Gardiner, professor of palaeontology at King's College, University of London, to include numerous bones and teeth, which showed clear signs of having been stained with potassium dichromate.

1912
Pharmacology
THE LONG ROAD TO ECSTASY

Anton Köllisch, a chemist working for the Merck pharmaceutical company, synthesized a new compound to rival hydrastinine, a drug patented by Bayer that stopped abnormal bleeding. In the course of his research, Köllisch synthesized methylenedioxymethamphetamine, which could be converted in turn to methylhydrastinine, a substance with the desired therapeutic effect. Methylenedioxymethamphetamine itself was largely forgotten until the 1970s, when a number of psychotherapists began to use the substance

– MDMA for short – to overcome the inhibitions of their patients. By the following decade, MDMA had metamorphosed into the core recreational drug of rave culture: Ecstasy.

TESTICLE TRANSPLANTS

Victor Lespinasse, professor of genitourinary surgery at Northwestern University, Illinois, reported that he had transplanted slices of human testicle into a man who had lost his own. Such was the success of this treatment, Lespinasse claimed, that the man insisted on leaving hospital early to make use of his newly restored capabilities. The following year Dr Frank Lydston transplanted the testicle of a suicide into his own scrotum, while in 1922 Dr L.L. Stanley, resident physician at California's San Quentin jail, reported that he had successfully transplanted the testicles of executed convicts. He went on to inject over 600 men with liquidized gonads of billy goat, ram, stag and boar. Belief in the rejuvenating effects of such procedures continued for many more years: it was only in 1997 that the German government banned the injection of tissue from sheep foetuses into the buttocks of patients who wished to recover the vigour of youth.

BOHR'S BOSH

Einstein famously expressed his dislike of Niels Bohr's quantum theory, saying 'God does not play dice with the universe.' Bohr is said to have retorted, 'Who are you to tell God what to do?' In 1913 two of Einstein's associates, Otto Stern and Max von Laue, while walking together up a mountain near Zurich, shook hands and swore that 'If this nonsense of Bohr's should in the end prove right we will leave physics.' Of course, they broke their oath: von Laue won the 1914 Nobel prize for physics, while

Stern went on to win it in 1943. Decades later, Stephen Hawking remarked, 'God sometimes throws dice where they can't be seen.'

EINSTEIN'S EXCESSES

The Austrian physicist and philosopher Ernst Mach, celebrated for his work on airflow and shock waves, rejected the physics of the younger generation. 'I can accept the theory of relativity,' he wrote in 1913, 'as little as I can accept the existence of atoms and other such dogma.' Mach, who held that only sensations are real, regarded atoms as mystical entities, on the grounds that their existence could only be inferred, not detected.

THE DISAPPEARANCE OF RUDOLPH DIESEL

(29 September) Rudolph Diesel, inventor of the internal combustion engine named after him, boarded the SS *Dresden* at Antwerp, Belgium, and set sail for Harwich, England. When the ship docked the following morning, Diesel's cabin was found to be empty. The coverlet on his bed had been turned down, and Diesel's watch was hanging beside it, but the bed had not been slept in. A search of the vessel found no sign of Dr Diesel, but his landing ticket was still in his cabin, so he could not have disembarked. The ostensible purpose of his visit to England was to attend a meeting of the Consolidated Diesel Engine Manufacturers. However, he also had an appointment with the British Admiralty. This was at the time of a fierce naval arms race between Britain and Germany. One of the latter's key strategic weapons during the Great War that broke out less than a year later was to be its fleet of submarines, used to deny Britain imports of food and raw materials. And these submarines were

powered by diesel engines. There is no doubt that the German authorities would have been alarmed at the idea of Rudolph Diesel talking to a potential enemy. But no evidence of an assassination has ever come to light, and the disappearance remains a mystery.

A CRITICAL SHORTAGE OF LEECHES

Once the mainstay of the medical profession, leeches had been used for centuries to suck 'excess' blood from the patient and so re-balance the supposed 'four humours'. But by the early 20th century they were used far less frequently. However, they still had their devotees, and the serious shortage of leeches caused by the First World War moved Sir Arthur Everett Shipley, the Master of Christ's College, Cambridge, and an expert on parasitic worms, to write to *The Times*:

> Sir,
> Our Country has been for many months suffering from a serious shortage of leeches. As long ago as last November there were only a few dozen left in London, and they were second-hand. Whilst General Joffre, General von Kluck, General von Hindenburg, and the Grand Duke Nicholas persist in fighting over some of the best leech-areas in Europe, possibly unwittingly, this shortage will continue . . .

FULL NETTLE JACKET

With supplies of cotton unavailable due to the British naval blockade, the Germans started to make their military uniforms out of nettles. They used a mix of 85 per cent common stinging nettle and 15 per cent *Boehmeria nivea*, a tropical member of the nettle family.

A NEW USE FOR CONKERS

British children were encouraged to collect horse chestnuts, not for playground games, but for the production of acetone, used in the manufacture of cordite, the explosive used to propel bullets and shells.

MADNESS, MURDER AND MATHEMATICS

(17 November) The French mathematician André Bloch – who gave his name to Bloch's theorem and Bloch's constant – stabbed his brother Georges (also a noted mathematician) and his aunt and uncle to death. A serving artillery officer, Bloch was at the time of the killings on convalescent leave from the army, having suffered a fall from an observation post. He later explained that his act had been entirely rational: his motive was a eugenic one, he said, his intention being to eliminate those members of the family affected by mental illness. He was confined to the asylum at Charenton in the outskirts of Paris, where the Marquis de Sade had also spent his last years. Thereafter Bloch produced a string of papers, and corresponded with a number of renowned mathematicians, usually dating his letters 1 April. Few of his correspondents were aware of his circumstances.

CIGARETTES ESSENTIAL FOR MEN'S WELLBEING

General John J. Pershing declared that tobacco was 'as indispensable as the daily ration', and hence his troops were supplied with as many free cigarettes as they could smoke. The 20th century's great epidemic of lung cancer began some two decades later.

PINK FOR A BOY

The *Ladies' Home Journal* told its readers:

> There has been a great diversity of opinion on the subject, but the generally accepted rule is pink for the boy and blue for the girl. The reason is that pink being a more decided and stronger colour is more suitable for the boy, while blue, which is more delicate and dainty, is prettier for the girl.

The American *Sunday Sentinel* had in 1914 advised mothers:

> If you like the colour note on the little one's garments, use pink for the boy and blue for the girl, if you are a follower of convention.

'A CURE FOR THE LIVING DEAD'

Under this slogan, William J.A. Bailey of Bailey Radium Laboratories Inc., New Jersey, marketed Radithor, a universal tonic and cure-all that would

RADIUM v. GREY HAIR

Who'd Dream she was 50?

50—and not a grey hair to be seen. Wonderful ! Yet an absolute fact. Let 'CARADIUM' do for you what it has done for thousands of our clients in all parts of the world. 'CARADIUM' will quickly restore, right from the hair roots, the natural colour, health and beauty to your hair, making you look 10 to 20 years younger.

Write for Free Hair Book.

'Caradium' is NOT A DYE

CONTAINING RADIO-ACTIVE WATER

Regular application of 'Caradium' will revivify the colour glands of the hair and cause the natural pigment to flow afresh. 'Caradium' Restorer is just as efficacious in cases of premature or inherited greyness or greyness caused by illness, worry, or overwork. It is absolutely sure. So natural is the course of restoration, that the use of 'Caradium' is absolutely undetectable.

Grey Hair will never appear if CARADIUM IS USED ONCE WEEKLY AS A TONIC

Caradium Shampoo Powders (for dry or Greasy hair) are the finest in the world for producing Soft and Glossy hair, 6d. each, Packets of twelve, 5/-.

WARNING.—Ask for Caradium Regd. and see that you get it; imitations are useless.

Caradium REGD.

4/- A 4/- size is now available for those only slightly grey. 'Caradium' Hair Restorer is obtainable of all good Chemists, Harrods, Whiteleys, Barkers, Selfridge's, Timothy Whites, Boots, Taylor's Drug Stores, etc., or direct in plain wrapper, POST FREE U.K. (overseas postage 2/6 extra) from :— **Large Size 7/6**

'CARADIUM' REGD., 38 Great Smith Street, Westminster, London.

In the earlier 20th century, radioactive 'tonics' were claimed to help with a wide variety of conditions, from grey hair to impotence. This advertisement dates from 1934.

deliver 'perpetual sunshine'. Radithor comprised distilled water with doses of radium 226 and 228 – both highly radioactive. At this stage, and for some years more, radioactivity was thought to have beneficial effects. Radithor was eventually withdrawn after the American socialite Eben Byers died in 1932, having consuming 1400 bottles of the stuff and lost most of his jaw. His body was so radioactive, it was buried in a lead-lined coffin. (Bailey himself claimed to have drunk more Radithor than any living man. He died of bladder cancer in 1949.)

Despite the uproar following Byers's death, in the early 1930s Tho-Radia face cream was extremely popular in France. It contained 0.5 g thorium chloride and 0.25 mg radium bromide per 100 g, and the manufacturers claimed that it had been developed by a 'Dr Alfred Curie' (thought to be a fictional character rather than a member of the famous Nobel-winning family). In the war years of the following decade, a company in Berlin produced Doramad Radioactive Toothpaste, claiming that radioactivity strengthened the defences of the teeth and gums against bacteria, while polishing and whitening the teeth. Luckily for those who used it, the relatively small amounts of thorium in the toothpaste produced only very low levels of radioactivity. Other products marketed on the basis of their radioactivity included Radium Schokolade, a chocolate manufactured by the German firm of Burk & Braun between 1931 and 1936, and Vita Radium Suppositories, 'for rectal use by men', marketed by the Home Products Company of Denver, Colorado. The company explained that their suppositories were 'tone restorers of sex and energizers for the entire nervous, glandular and circulatory systems', and were particularly 'recommended for sexually weak men'. The suppositories, in which the radium was 'carried in a cocoa butter base', were also supposedly 'splendid for piles and rectal sores'. They were, of course, 'guaranteed entirely harmless'.

NOT A BORING NUMBER

(26 April) Death of the largely self-taught Indian mathematical genius Srinivasa Ramanujan, of whom it was said that every positive integer was one of his personal friends. He had been invited to come to Cambridge University by Professor G.H. Hardy, but neither the climate nor English food suited Ramanujan, and he succumbed to tuberculosis, perhaps hastened by malnutrition. As he lay dying, Hardy went to visit him in Putney. 'I had ridden in a taxi-cab No. 1729,' Hardy recalled, 'and remarked that the number seemed to me a rather dull one, and that I hoped it was not an unfavourable omen.' 'No,' Ramanujan replied, 'it is a very interesting number; it is the smallest number expressible as the sum of two cubes in two different ways.' The two ways are $1^3 + 12^3$ and $9^3 + 10^3$.

THE AWESOME ATAMAN

After the Red Army occupied Odessa during the Russian Civil War, Igor Tamm – who went on to win the Nobel prize for physics in 1958 – left the city to try to trade silver spoons for chickens. He was captured by one of the roving Cossack bands who made up the Ukrainian anarchist army of Nestor Makhno, who fought both Reds and Whites. His captors, suspicious of his city clothes, took him to their Ataman (leader), a fearsome looking figure sporting a wild beard, fur hat and bandoliers of cartridges and hand grenades. 'You communist bastard, working against our mother Ukraine!' shouted the Ataman. 'The sentence is death.' Tamm protested that he was just a humble professor of mathematics at the University of Odessa, and was in the countryside searching for food. 'A maths professor?' shouted the Ataman. 'If you're a maths professor then give me an estimate of the error one makes by cutting off Maclaurin's series at the nth term. If you can't, I'll have you shot.' The terrified Tamm was astonished that the Ataman should be familiar with this abstruse branch of mathematics, and with the guns of his captors trained on him he rather shakily worked out the solution. 'Correct!' shouted the Ataman. 'Now run along home.' The identity of this mysterious Ataman has never been established.

EDISON'S SPIRIT COMMUNICATOR

Thomas Edison, known for a myriad inventions including the phonograph, the movie camera and the first practical light bulb, outlined a device to 'furnish psychic investigators with an apparatus which will give a scientific aspect to their work'. He continued:

> This apparatus, let me explain, is in the nature of a valve, so to speak. That is to say, the slightest conceivable effort is made to exert many times its initial power for indicative purposes. It is similar to a modern power house, where man, with his relatively puny one-eighth horse-power, turns a valve which starts a 50,000-horse-power steam turbine. My apparatus is along those lines, in that the slightest effort which it intercepts will be magnified many times so as to give us whatever form of record we desire for the purpose of investigation. Beyond that I don't care to say anything further regarding its nature. I have been working out the details for some time; indeed, a collaborator in this work died only the other day. In that he knew exactly what I am after in this work, I believe he ought to be the first to use it if he is able to do so.

Opinion is divided as to whether Edison was just pulling the collective leg of the spiritualists. Elsewhere, in a private letter quoted by his biographer Paul Israel, he stated that 'It is doubtful in my opinion if our intelligence or soul or whatever one may call it lives hereafter as an entity or disperses back again from whence it came, scattered amongst the cells of which we are made.'

SILENCE FROM THE FAR SIDE

In the interests of science, a Detroit electrical engineer called Thomas Lynn Bradford committed suicide, having previously instructed his associate, a spiritualist medium called Ruth Doran, to record his reports of life after death to the scientific community. With admirable honesty, Miss Doran admitted that after Bradford gassed himself she had received no communications whatsoever from the deceased.

BESPOKE CHICKEN BITS

In the magazine *Popular Mechanics*, Winston Churchill predicted that in 50 years' time 'we shall escape the absurdity of growing a whole chicken in order to eat the breast or wing, by growing those parts separately under a suitable medium'.

NOBEL LAUREATES AT ODDS

Einstein was awarded the Nobel prize for physics – not for his theories of special and general relativity, but for his quantum explanation of the photoelectric effect. The photoelectric effect had been discovered by Philipp Lenard, who was awarded the 1905 Nobel prize as a consequence. Ironically, Lenard, a virulent German nationalist and anti-Semite, never accepted Einstein's explanation, and denounced relativity as a 'Jewish fraud'. During the First World War he wrote to James Franck, a future Nobel laureate then fighting on the Western Front, urging him to put his all into defeating the British – who, he complained, had never cited his works correctly. Lenard went on to become an ardent Nazi, and under Hitler was appointed to the post of 'Chief of Aryan Physics'. In 1945 the Allied occupying forces ejected him from his post as professor emeritus at the University of Heidelberg. He died two years later, aged 84.

GENIUS OR IGNORANT NUMBSKULL?

Werner Heisenberg, pioneer of quantum mechanics, author of the uncertainty principle and future Nobel laureate, was subjected to a gruelling oral examination at the University of Munich for his doctoral degree. He was expected to be knowledgeable about both theoretical and experimental physics, but the latter was not his forte, and he was unable to answer questions about the resolving power of a telescope or a microscope, and could not explain the operation of a lead storage battery. One of his examiners, Arnold Sommerfeld, recognized Heisenberg's astonishing theoretical abilities, and in his report spoke of the young man's 'unique genius'. But the other examiner, Wilhelm Wien, who had won the 1911 Nobel prize for his experimental work on heat radiation, was far from impressed. Heisenberg, he said, suffered from 'bottomless ignorance'.

1923

Physiognomy / psychology

JUST OBEYING ORDERS – PART I

Carney Landis, a researcher at the University of Minnesota, made a photographic record of people's facial expressions while they were subjected to various stimuli – such as listening to music, smelling ammonia, perusing erotica or immersing their hands in a bucketful of frogs. The final part of the experiment came when he asked them to cut off the head of a rat with a knife. Most were reluctant to do so, but in the end complied – anticipating the notorious 'Milgram Experiment' (*see* 1961). However, Landis did not pick up on this by-product of his research, merely noting that people exhibited a variety of expressions while decapitating a rat.

1925 to 1949

A lunatic in reception ✱ The case of
the midwife toad ✱ Towards an ape-
human hybrid? ✱ The world's slowest
experiment ✱ Ignose and godnose
✱ Planet inadvertently named after
laxative ✱ Goat fails to transmute into
man ✱ Reviving the dead ✱ Decapitation
encourages copulation ✱ Alcohol on the
brain ✱ A request for severed testicles
✱ The fur-bearing trout ✱ Lunar insects
and other follies ✱ Birth of the googol
✱ On the uses of a perforated eardrum
✱ Stormtroopers on speed ✱ On the
benign effects of mustard gas ✱ Safe
cracking, plate juggling and quantum
electrodynamics ✱ The people's war
against the snail

A LUNATIC IN RECEPTION

When John Logie Baird arrived at the offices of the *Daily Express* to demonstrate his television system, the editor refused to see him. 'For God's sake go down to reception,' he told a minion, 'and get rid of the lunatic who's down there. He says he's got a machine for seeing by wireless! Watch him – he may have a razor on him.' Baird went on to organize a public demonstration instead. The following year the electronics pioneer Lee De Forest, although conceding the technical viability of television, doubted it had any commercial future. It was, he said, 'a development of which we need waste little time dreaming'. Ten years later, in 1936, the editor of the

John Logie Baird adjusting a transmitter in his pioneering electro-mechanical system of 'wireless vision'. The editor of the Daily Express *thought he was a dangerous lunatic.*

Radio Times, Rex Lambert, opined that 'Television won't matter in your lifetime or mine,' while ten years after *that*, in 1946, the Hollywood film producer Darryl F. Zanuck, complacently commented, 'People will soon get tired of staring at a plywood box every night.'

NIGHT FALLS ON TENNESSEE

The Tennessee state legislature passed an act, drafted by a farmer called John Washington Butler, that included the following provisions:

> It shall be unlawful for any teacher in any of the universities, normals and all other public schools of the State which are supported in whole or in part by the public school funds of the State, to teach any theory that denies the Story of the Divine Creation of man as taught in the Bible, and to teach instead that man has descended from a lower order of animals.

That same year a teacher called John Scopes was successfully prosecuted under the act in the famous 'Monkey Trial', and fined $100. On appeal, the Tennessee Supreme Court did not accept any of the defence's arguments, but they did set aside the guilty verdict on a legal technicality. The Butler Act was only repealed in 1967.

THE CASE OF THE MIDWIFE TOAD

The Austrian biologist Paul Kammerer was a follower of the pre-Darwinian evolutionary doctrine of Jean-Baptiste Lamarck (*see* 1809), who held that acquired characteristics could be inherited – hence the ancestors of giraffes who stretched their necks longer to snack on ever higher leaves passed these longer necks onto their offspring. By Kammerer's day, Lamarckism had been swept aside by Darwin's theory of natural selection, but Kammerer was determined to prove Darwin wrong, and Lamarck right. He chose as his subject the midwife toad, which, unlike other toads, breeds on dry land, and has thus dispensed with the nuptial pads – black scaly bumps on the hind legs – that the males of other species of toad use to grip onto the females while mating in water.

Kammerer set about making his midwife toads breed in water by increasing the temperature, and over the generations he claimed that his toads were developing black nuptial pads. When Dr G.K. Noble, curator of reptiles at the American Museum of Natural History, examined some of Kammerer's specimens he concluded that the black nuptial pads were not nuptial pads at all. Rather, someone had injected Indian ink under the toads' skin. Six weeks after Noble published his findings in *Nature*, Kammerer walked out into the forest of Schneeberg and shot himself. It has been suggested that it was not Kammerer who had tampered with the specimens, but rather a Nazi sympathizer, who hated Kammerer for his pacifism and his socialism, and who wished to discredit him.

Male midwife toad with chains of eggs.

USING ONE DISEASE TO TREAT ANOTHER

The Austrian neurologist Julius von Wagner-Jauregg was awarded the Nobel prize for his bizarre method of treating patients with 'general paralysis of the insane' (GPI), the late stage of syphilis. Noting that such patients showed some improvement in their mental state after suffering a fever, and that cultures of *Treponema pallidum*, the bacteria responsible for syphilis, could be killed off if the test tube was heated, he decided on a radical experiment. He deliberately infected some of his patients with malaria, which produces high fevers, and indeed many of them appeared to show a clear improvement – or at least a retardation in the development of dementia. He then treated the malaria with quinine. His treatment caught on, and around the world thousands of patients with GPI were deliberately given malaria. The treatment was only abandoned once penicillin came into use.

TOWARDS AN APE-HUMAN HYBRID?

(28 February) At the botanical gardens in Conakry, French Guinea, the Soviet scientist Ilya Ivanovich Ivanov began a series of experiments to create an ape-human hybrid by artificially inseminating two female chimpanzees with human sperm. In June, he repeated the experiment with a third female chimpanzee, but no signs of pregnancy occurred. He also tried to arrange for the insemination of human females with chimpanzee sperm, but the French authorities were not happy with such a prospect. Returning to the Soviet Union later that year, Ivanov gained the support of the Society of

Materialist Biologists to carry out his plan, using five women volunteers. However, in June 1929, before he could proceed, he learnt that his last mature male ape, an orang-utan, had died. The following year, Ivanov fell out of political favour and was exiled to Kazakhstan, where he died in 1932. His obituary was written by Ivan Pavlov. Geneticists have since established that humans and chimpanzees have 95 per cent of their DNA in common, and that the two lineages only diverged some 6 million years ago – and may have continued to interbreed for a further 1.2 million years.

THE WORLD'S SLOWEST EXPERIMENT

The so-called pitch-drop experiment began at the University of Queensland, Australia, initiated by Professor Thomas Parnell. A mass of congealed black tar pitch was placed in a funnel to see how fast it would flow. The experiment is still in progress, and on average one drop of pitch drips through the funnel every nine years – although no one has ever actually witnessed this happen. The conclusion is that this pitch has a viscosity approximately 230 billion times that of water.

THE CONSEQUENCES OF DROPPING ANIMALS DOWN MINESHAFTS

In his essay 'On Being the Right Size', the biologist and science popularizer J.B.S. Haldane described the effects of gravity on various sizes of animal: 'You can drop a mouse down a thousand-yard [900 m] mine shaft; and, on arriving at the bottom, it gets a slight shock and walks away. A rat is killed, a man is broken, a horse splashes.' But tiny animals such as insects face other dangers that humans do not, he pointed out – for example, surface tension. 'A man coming out of a bath carries with him a film of water of about one-fiftieth of an inch [0.5 mm] in thickness. This weighs roughly a pound [0.45 kg]. A wet mouse has to carry about its own weight in water. A wet fly has to lift many times its own weight and, as everyone knows, a fly once wetted by water or any other liquid is in a very serious position indeed.'

IGNOSE AND GODNOSE

The Hungarian scientist Albert von Szent-Györgyi discovered a hitherto unknown compound in cabbages, oranges and the tissue of the adrenal gland. Suspecting it was some kind of sugar, akin to glucose, fructose, lactose, etc., he named the mysterious substance 'ignose' – reflecting his own ignorance. When he submitted a paper on his discovery to the *Biochemical Journal*, the editor requested that Szent-Györgyi rename the compound. Szent-Györgyi duly responded with a new name: 'godnose'. Exasperated, the editor named the compound himself, calling it 'hexuronic acid', referring to its six carbon atoms. The chemical is now known as 'ascorbic acid' – or, more familiarly, as 'vitamin C'. Szent-Györgyi was awarded a Nobel prize in 1937.

SMELLING A RAT

Ernest Lawrence built the first cyclotron, a type of particle accelerator, at the University of California, Berkeley. One of the experiments conducted using the cyclotron involved directing 'neutron rays' at a rat confined in a small cylinder. The experimenters had no idea how large a dose would achieve an observable effect, so arbitrarily limited the exposure to two minutes. On opening the cylinder, the rat was found to be dead.

All involved found this a sobering warning of the dangers of nuclear radiation, and as a consequence both scientists and technicians were meticulous in their precautions while operating the cyclotron. So as not to diminish this safety-conscious attitude, those who carried out a post-mortem on the rat and found that it had died not from the effects of radiation but from asphyxiation in the tight confines of the cylinder did not advertise this verdict widely.

THE WORLD TURNED TOPSY-TURVY

In the wake of the revelations of quantum physics, F.W. Bridgman, professor of mathematics and philosophy at Harvard, declared in *Harper's Magazine* that 'This means that nothing more nor less than that the law of cause and effect must be given up . . . The world is not a world of reason, understandable by the intellect of man.' In the light of this, not only would we have to develop new methods of teaching young children, Bridgman argued, we would also need to remodel our languages to extirpate all forms of speech that assume that cause leads inexorably to effect.

PLANET INADVERTENTLY NAMED AFTER LAXATIVE

Following the discovery by Clyde Tombaugh of a new planet far beyond the orbit of Neptune, hundreds of suggestions poured in from all round the world as to what it should be called. Eventually the members of the Lowell Observatory in Arizona, where the discovery had been made, came up with a shortlist of three: Minerva, Cronos and Pluto. Pluto had been suggested by Venetia Burnley, an 11-year-old schoolgirl from Oxford, England, who thought the name of the Greek god of the Underworld suitable for such a dark and cold place. The members duly voted in favour of Pluto, disregarding the fact that to most Americans at the time the name was primarily associated with Pluto Water, a brand of laxative marketed under the slogan 'When nature won't, Pluto will.' In 2006 Pluto was demoted to the status of 'dwarf planet'.

STUFF AND NONSENSE – PART II

In his book *The Philosophy of Biology*, the distinguished biologist J.S. Haldane (father of the geneticist J.B.S. Haldane) declared that it was 'inconceivable' that heredity should be passed on by a molecule. Just over two decades later, Francis Crick and James Watson described the structure of DNA.

FAMOUS LAST WORDS

The distinguished British abdominal surgeon Lord Moynihan declared, 'We can surely never hope to see the craft of surgery made much more perfect than it is today.' Despite his apparent complacency, Moynihan had pioneered the techniques of aseptic surgery, involving scrubbing up, cap, gown and rubber gloves. Even specialized footwear was required, drawing mockery from some quarters. When a French surgeon of the old school saw Moynihan in his rubber shoes prior to an operation, he inquired, 'Surely he does not intend to *stand* in the abdomen?'

STATIC FROM THE STARS

Working at the Bell Telephone Laboratories at Holmdel, New Jersey, Karl Jansky was tasked with detecting sources of static that might interfere with the company's planned transatlantic telephone service. To this end he built a sensitive antenna mounted on a turntable (dubbed 'Jansky's Merry-go-round'), and with it identified three sources of interference: nearby thunderstorms, distant thunderstorms, and a faint hiss of mysterious origin. He monitored this source over months, and noted that it varied in intensity over a period of 23 hours and 56 minutes – the length of the sidereal day. The sidereal day – as opposed to the solar day of 24 hours – is the period of the Earth's rotation relative to the cosmos, so Jansky concluded that the radio source was not the Sun, as he had first suspected, but must be far further away. Eventually he traced the source to the constellation of

Karl Jansky with his famous 'Merry-go-round', an unlikely looking contraption for detecting hitherto unsuspected radio signals from outer space.

Sagittarius, towards the centre of the Milky Way. Jansky asked Bell for more funding so that he could investigate further, but Bell turned him down, on the grounds that these distant emissions would not adversely affect the transatlantic transmissions. Thus the man who founded radio astronomy was unable to undertake any more work in the field.

GOAT FAILS TO TRANSMUTE INTO MAN

Harry Price, founder of the National Laboratory for Psychical Research, travelled to Germany in order to turn a goat into a young man by means of black magic. By all accounts, the experiment – which was intended to celebrate the centenary of Goethe's death – was not a success.

THE TUSKEGEE SYPHILIS STUDY

In Tuskegee, Macon County, Alabama, the US Public Health Service began one of the most unethical experiments ever undertaken. Nearly 400 impoverished African American sharecroppers suffering from syphilis were identified, and their symptoms monitored over the following decades. Despite the fact that cures were available – particularly after the introduction of penicillin in the 1940s – the subjects were not informed of these, and were left untreated. As a result, many of the subjects died, and many of their wives and children were infected. The study was only abandoned in 1972, after a leak to the press. In a 1976 interview, Dr John Heller, director of venereal diseases at the Public Health Service between 1943 and 1948, stated: 'The men's status did not warrant ethical debate. They were subjects, not patients; clinical material, not sick people.'

ON THE BENEFITS OF FACIAL HAIR

Governor William 'Alfalfa Bill' Murray of Oklahoma declared that 'It's a scientific fact that if you shave your moustache, you weaken your eyes.'

'THE WORLD WAS HEADED FOR GRIEF'

The Hungarian-Jewish scientist Leó Szilárd arrived in London, having fled Nazi persecution in Germany, where he had worked at the University of Berlin. One morning that autumn he read in *The Times* that Lord Rutherford had declared that anyone who proposed that atomic energy could be liberated on an industrial scale was 'talking moonshine'. 'Pronouncements of experts to the effect that something cannot be done have always irritated me,' he later wrote, and went on to describe how, later that day while waiting for traffic lights to change on Southampton Row in Bloomsbury, he conceived of the idea of the nuclear chain reaction. Realizing that this could become the basis of an atomic bomb, he went on to take out a patent, which in 1936 he assigned to the British Admiralty, to keep the idea secret. In 1938 Szilárd moved to the USA, where he worked with Enrico Fermi at Columbia University. He had tried to initiate chain reactions using the elements beryllium and indium, with no success, but in 1939 he and Fermi heard of the successful nuclear fission experiment conducted in Germany by Otto Hahn, Fritz Strassmann and Lise Meitner. He and Fermi realized that uranium had the potential to undergo a chain reaction, and set about putting it to the test. 'We turned the switch and saw the flashes. We watched them for a little while and then we switched everything off and went home.' Szilárd fully realized the implications. 'That night, there was very little doubt in my mind that the world was headed for grief.' Realizing that the Nazis were already working on nuclear weapons, Szilárd asked his old friend and collaborator Albert Einstein to join him in writing to President Franklin D. Roosevelt, urging him to embark on what became the Manhattan Project. Once the bomb was developed, in July 1945 Szilárd organized a petition signed by dozens of scientists to ask President Harry S. Truman to consider demonstrating the power of the new weapon to the Japanese, before using it on people. The petition never reached the president.

REVIVING THE DEAD

Dr Robert E. Cornish of the University of California at Berkeley embarked on a series of experiments to see whether he could revive dead dogs. A child prodigy, Cornish had been awarded his doctorate at the age of just 22, and had interested himself in a number of scientific projects, such as devising a method of reading newspapers underwater and bringing dead people back to life. In pursuit of the latter, in 1933 he tried to revive three men several hours after they had suffered fatal heart attacks. His technique involved placing them on a 'teeterboard' – a sort of see-saw – and moving them rapidly up and down, to try to get the circulation re-started, while at the same time delivering oxygen via a mask. His efforts were unsuccessful, so to perfect his technique Cornish turned to experimenting on fox terriers. His subjects, all called Lazarus, were asphyxiated with ether and nitrogen, and after some minutes the attempt at revival began. The dead dog was placed on the teeterboard, and a drip was inserted into a vein in its leg, through which was fed saline solution saturated with oxygen, adrenaline and anticoagulants. While Cornish 'breathed gustily' into the dog's mouth, his assistant vigorously rubbed the dog's body while rocking it up and down on the teeterboard. In a report dated 26 March 1934, *Time* magazine described what happened next, in the case of Lazarus II:

The stimulant solution sank in a glass gauge as it seeped into the corpse through five feet of rubber tubing. In a little while the gauge level stopped falling, began to rise in slow pulsations. Lazarus II gasped. His leg twitched. His heart began to beat, feebly at first, then like a trip-hammer, then normally. Lazarus II was alive. For eight hours and 13 minutes the dog lay in an uneasy coma, whining, panting, barking, as if ridden by nightmares. Eager to speed recovery, Dr Cornish injected some glucose solution. A blood clot formed and Lazarus II died again, this time for good and all.

Cornish – who played himself in a 1934 Universal feature film called *Life Returns* – had more success with Lazarus IV. Lazarus IV lived for several months after resuscitation, and, although blind and unable to stand unaided, learned, according to *Modern Mechanix* (January 1935), 'to crawl, bark, sit up on its haunches and consume nearly a pound of meat a day'.

Cornish's experiments proved too controversial for Berkeley, and he was obliged to continue his work at his own home. In 1947 he secured the agreement of a man on death row to be a guinea pig in a post-execution revival effort. But the authorities vetoed the suggestion, presumably on the grounds that the experiment, if successful, would make a mockery of the whole idea of capital punishment.

AN ADDICTION TO CHESS

(17 October) Death of the Spanish scientist Santiago Ramón y Cajal, who with Camillo Golgi won the 1906 Nobel prize for showing that the neuron or nerve cell was the basic unit of the nervous system. In his memoirs he recalls how in his youth his scientific career was almost sacrificed on the altar of chess, a game that inspired him with 'a morbid eagerness to overcome my adversaries'. As his obsession began to take over, at the local chess club he

would show off by playing four opponents simultaneously, and even played without looking at the board. He would devour every book he could find on the game, send off solutions to chess problems in foreign magazines, and even found his sleep broken 'by dreams and nightmares, in which pawns, knights, queens and bishops were jumbled together in a frenzied dance'. He concluded that things could not go on like this:

> The almost permanent fatigue and cerebral congestion weakened me. If one does not lose money in playing chess, one loses time and brain energy, which are worth infinitely more, and one's will is turned aside and runs through the wrong channels. In my opinion, far from exercising the intelligence, as many claim, chess warps it and wears it out.

He realized that he could not just stop, so adopted a cunning stratagem to placate his ego. Abandoning his usual style of play, which involved 'romantic and audacious attacks', he decided to 'stick to the rules of most cautious prudence'. Following this course, he succeeded in defeating each and every one of his opponents in the course of a week, and thus 'the devil of pride smiled and was satisfied'. After that, he did not touch a pawn again for a quarter of a century, and was able to devote his intellect, fully and without distraction, 'to the noble worship of science'.

DECAPITATION ENCOURAGES COPULATION

In his paper 'An Experimental Analysis of the Sexual Behaviour of the Praying Mantis' (The *Biological Bulletin* 69, October 1935), K.D. Roeder of Tufts College, Massachusetts, reported that there appeared to be an evolutionary advantage in the long-observed habit of female mantises devouring their mates during the sexual act:

Since his head is attacked first, an inhibitory centre in the suboesophageal ganglion is soon destroyed. This centre normally inhibits (1) lateral locomotor movements, (2) copulatory movements of the abdomen, which originate in the last abdominal ganglion. These movements consequently commence after destruction of the head and bring the body of the male into the mating position on the back of the female, and copulation is immediately effected.

Roeder found that if he decapitated a male, it would – after it got over the shock (a period of some ten minutes) – violently attempt to copulate with any long, rounded object, such as a pencil or a finger. Roeder also found that 'A decapitated female will readily accept a male, decapitated or otherwise . . . The pair remain together for about four hours.'

ALCOHOL ON THE BRAIN

Following up on earlier experiments, the Portuguese physician António Egas Moniz and a surgical colleague pioneered a treatment for schizophrenia involving drilling holes in the patient's skull and then pouring ethyl alcohol into the prefrontal cortex. By disrupting the neuronal tracts, Moniz hoped to reduce his patient's extreme anxiety and paranoia, and this indeed seemed to be the result. Moniz (who was awarded a Nobel prize in 1949) went on to develop an instrument that would have the same effect by removing segments of white brain matter, although the results of subsequent operations were variable.

This operation became known as prefrontal leucotomy or lobotomy, and although its use was opposed by many neurosurgeons, it was strongly advocated by two American neurologists, Walter Freeman and James Watts. As a result, many people with mental disorders, or with mentally ill relatives, clamoured for this 'miracle cure'. Freeman went on to develop a technique in the 1940s that did not require large-scale open-skull surgery, and which could be performed in a matter of minutes. This involved inserting a pick

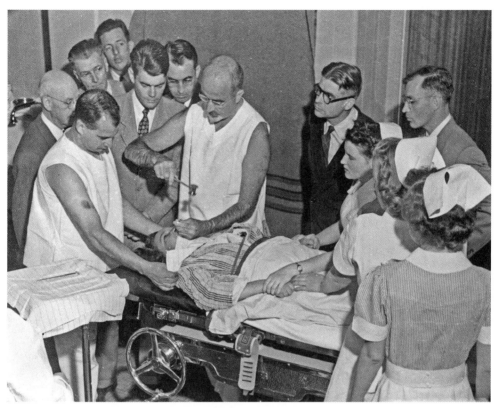

Dr Freeman demonstrating his method of lobotomizing a patient.
This involved inserting a pick into the front of the brain through the eye socket.

through the back of the eye socket and using it to sever the connections between the frontal lobes.

This procedure was carried out on tens of thousands of patients, some with only minor mental disorders, and although there were reductions in symptoms such as anxiety, many suffered side effects such as apathy, loss of concentration, and a diminished ability to experience emotion. One of the most notorious cases was that of the sister of the future US president John F. Kennedy, Rosemary Kennedy, whose father arranged for her to have a lobotomy at the age of 23 to cure her 'moodiness'; as a consequence, Rosemary's mental age was reduced to that of an infant.

In his 1948 book *Cybernetics*, Norbert Weiner trenchantly noted:

> Prefrontal lobotomy . . . has recently been having a certain vogue, probably not unconnected with the fact that it makes the custodial care of many patients easier. Let me remark in passing that killing them makes their custodial care still easier.

Lobotomies continued to be carried out into the 1970s, but from the 1950s the introduction of new drugs began to make surgery a less attractive option.

A REQUEST FOR SEVERED TESTICLES

Following the Italian invasion of Abyssinia (now Ethiopia), the Polish biochemist Casimir Funk – known for his pioneering work on vitamins and sex hormones – sent a peculiar request to the Abyssinian government. He had heard, he said, that Abyssinian tribesmen were in the habit of castrating any Italian soldiers they captured. If this were the case, he would be much obliged, he said, if the government of Abyssinia would forward to him any of these trophies, for the purposes of his scientific research. He was met by a diplomatic silence.

ON THE DANGERS OF TIDAL POWER

The distinguished colono-rectal surgeon John P. Lockhart-Mummery published *After Us*, a collection of essays that ranged beyond the confines of proctology. In one of these essays he predicted that in the future the tides might be used to generate power on a large scale. But there was a catch. 'If extensively exploited over a long period of time,' he warned, 'it might result in bringing the Moon too close to Earth for safety.'

DRUG KILLS CHILDREN WITH SORE THROATS

The US pharmaceutical firm S.E. Massengill produced a new liquid form of the drug sulphanilamide, used for treating streptococcal infections. Sulphanilamide was already available as a pill or a powder, but to make it more palatable for children, Massengill used diethylene glycol – which has a pleasant sweet taste – as a liquid medium, and sold the new product as 'Elixir Sulfanilamide'.

At this date, there was no legal requirement for new drugs to be subjected to toxicological testing, and it was not long before it became clear that this one was deadly. All those who took it suffered appalling symptoms: nausea, vomiting, urine retention, stupor, convulsions and agonizing abdominal pain. The mother of one of the dead children wrote to President Franklin D. Roosevelt: 'The first time I ever had occasion to call in a doctor for [Joan] and she was given Elixir of Sulfanilamide. All that is left to us is the caring for her little grave. Even the memory of her is mixed with sorrow for we can see her little body tossing to and fro and hear that little voice screaming with pain and it seems as though it would drive me insane.' In all, 107 people died of kidney failure, many of them children.

The chief chemist at Massengill, Harold Cole Watkins, committed suicide, but there was no law under which the company itself could be prosecuted, apart from the technical charge of 'misbranding' (the product should have been labelled a 'solution' rather than an 'elixir', as it contained no alcohol). The resultant nationwide outrage led to the passage of the Federal Food, Drug and Cosmetic Act of 1938.

Diethylene glycol – which is now used as a solvent and in antifreeze – has gone on killing people in other countries, where unscrupulous drug manufacturers have used it as a cheap alternative to pharmaceutical-grade glycerine or propylene glycol. In one incident alone in Bangladesh in the early 1990s over 300 children died.

In 1985 it was found that millions of bottles of Austrian wine contained up to 0.1 per cent diethylene glycol, which had been deliberately added to provide sweetness. Having seized vast quantities of the contaminated wine, the Austrian government was at a loss as to what to do with it, until it was found that when the wine was mixed with salt it provided an effective way to clear ice from the streets in winter.

THE FUR-BEARING TROUT

(15 November) The *Pueblo Chieftain* newspaper reported that 'Old-timers living along the Arkansas near Salida have told tales for many years of the fur-bearing trout indigenous to the waters of the Arkansas near there.' Tales of 'furried fish' in North American rivers go back to the 17th century. Some said that the fish grew the hair to maintain body heat in the cold northern waters, while others blamed a spillage of hair tonic into the Arkansas River. The origin of the tall tale may lie in the fungus *Saprolegnia*, which can infect fish and causes the growth of fur-like fruiting bodies on the outside of the body.

LUNAR INSECTS AND OTHER FOLLIES

(17 January) Death of the American astronomer William Henry Pickering, whose otherwise distinguished career was marred by a few wild speculations

and some rash predictions. For example, he claimed to have detected vegetation on the Moon, and suggested that changes in the appearance of the lunar crater Eratosthenes were caused by 'lunar insects'. Regarding the newly invented aeroplane he was, in contrast, extremely sceptical. 'A popular fallacy is to expect enormous speed to be obtained,' he wrote in 1908. 'There is no hope of competing for racing speed with either our locomotives or our automobiles.' He also dismissed notions of 'gigantic flying machines speeding across the Atlantic and carrying innumerable passengers in a way analogous to our modern steamships'. Another popular fantasy, he wrote, 'is to suppose that flying machines could be used to drop dynamite on the enemy in time of war'.

The craters called Pickering on both the Moon and Mars are named in honour of Pickering and his brother, also a distinguished astronomer.

BIRTH OF THE GOOGOL

The term *googol* was coined by Milton Sirotta, the nine-year-old nephew of the US mathematician Edward Kasner. A googol is 10^{100}, i.e. 10,000,000,000,000,000,000,000,000,000,000,000,000,000,000,000,000, 000,000,000,000,000,000,000,000,000,000,000,000,000,000,000,000.

The boy also proffered the term *googolplex* for a considerably vaster number: 1 followed by a googol of zeros.

THE WIND BAG

A patent application was lodged for the 'Wind Bag', designed 'for the receiving and storing of gas formed by the digestion of foods'. A tube linked the rectum to a collection chamber, while the device was held in place under one's clothes by a belt.

ON THE USES OF A PERFORATED EARDRUM

1939
Physiology

Following the sinking of the submarine HMS *Thetis* with the loss of 99 lives, the biologist J.B.S. Haldane was asked by the Admiralty to look into improving the methods of escape from crippled submarines. There followed a series of experiments in which Haldane himself took considerable risks, exposing himself to high pressures and high concentrations of carbon dioxide and oxygen: on one occasion, he breathed oxygen at a pressure of 4 atmospheres while immersed in a bath surrounded by blocks of ice. One effect of the pressure experiments was perforated eardrums, but Haldane shrugged this off: 'The drum generally heals up,' he later wrote, 'and if a hole remains in it, although one is somewhat deaf, one can blow tobacco smoke out of the ear in question, which is a social accomplishment.'

STORMTROOPERS ON SPEED

(May) As the German Panzer divisions swooped through the Ardennes towards the River Meuse, the tank crews of Army Group A were kept alert by some 20,000 tablets of methamphetamine. This was as nothing compared to the 200 million amphetamine tablets issued to US servicemen during the war, to fend off fatigue. During the war and into the 1950s around half a million industrial workers in Japan were routinely given methamphetamine to boost their productivity, with calamitous consequences for their health.

CRACKING THE MAUD CODE

(June) Two months after the Nazi occupation of Denmark, the Danish physicist Niels Bohr sent a cryptic telegram to the German physicist Otto Frisch, who was exiled in England and working on the very early stages of Britain's nuclear weapons programme. The telegram concluded with the words 'Tell Cockcroft and Maud Ray Kent'. John Cockcroft was the British Nobel laureate who had split the atomic nucleus in 1932, but who Maud Ray was no one knew. The assumption, therefore, was that 'Maud Ray Kent' was a cipher. Frisch recalled that he and his colleagues tried to see if they could work it out as an anagram, taking some freedom with the spellings. 'Radium taken' was one suggestion; 'U and D may react' another (U being uranium, and D being deuterium, a heavy isotope of hydrogen). Inspired by the telegram, the committee in charge of the British atomic project, on which Cockcroft served (but Frisch, as an enemy alien, could not), took the name 'Maud' – which conveniently stood for Military Application of Uranium Detonation. It was not until after the war that they found out that Maud Ray had once been governess to Bohr's children, and lived in Kent. The Maud Committee disbanded in July 1941, and thereafter Britain pooled its nuclear weapons research with the Americans, resulting in the Manhattan Project.

WHAT'S IN AN R?

Geoffrey Tandy, a curator at the Natural History Museum, was recruited to work at Bletchley Park, where thousands of mathematicians, translators and others were tasked with unravelling the top-secret German Enigma code. The officer who had taken on Tandy had thought he was an expert in *cryptograms*, whereas in fact he was an expert in *cryptogams*, plants such as algae, ferns and mosses that reproduce via spores rather than seeds. Tandy's expertise was nevertheless to come in useful when some waterlogged German notebooks containing coded messages were retrieved from a U-boat. To preserve them for the codebreakers, Tandy dried them out by laying the sheets between absorbent paper – just as in his previous career he had dried out specimens of seaweed.

COCONUTS IN THE BLOOD

During the fighting in the Pacific theatre, the Americans used coconut water, which is both sweet and sterile, in emergencies as a substitute for sterile glucose solution, and administered it intravenously to sick or injured men.

AN ABSENT-MINDED PROFESSOR

(14 February) Death of the renowned German mathematician David Hilbert, notorious for his absent-mindedness. On one occasion, when they were expecting guests for dinner, his wife told him to change his grubby tie before they arrived. Hilbert went off to his bedroom to change, but never reappeared. When a search was made, he was found slumbering in his bed. The prompt of undoing his tie had unleashed his normal bedtime routine, and he had undressed himself, put on his pyjamas, climbed into his bed and fallen fast asleep.

MISSING THE BOAT

Thomas J. Watson, chairman of IBM, reportedly stated: 'I think there is a world market for about five computers.' A quarter of a century later, in 1968, Robert Lloyd of IBM's Advanced Computing Systems Division was scornful of the new microprocessor. 'What the hell,' he asked, 'is it good for?'

ON THE BENIGN EFFECTS OF MUSTARD GAS

(3 December) A German air raid on the southern Italian port of Bari resulted in the sinking of 17 Allied vessels, and left several more badly damaged. Among those destroyed was the Liberty ship SS *John Harvey*, which was blown apart by a massive explosion, killing everybody on board. As the wounded were taken to hospital, many complained of hacking coughs and difficulties breathing. The doctors were baffled, until it was eventually realized that the *John Harvey* had been carrying a secret cargo of 2000 M47A1 mustard-gas bombs. Mustard gas was never used during the Second World War, but both sides kept stocks in reserve, just in case. In this case, the policy backfired badly: the accidental release of mustard gas in Bari led to 628 deaths among Allied servicemen and merchant seamen. There was one positive outcome, however. US army doctor Cornelius Rhoads, who treated some of the victims, noted that their levels of white blood cell – a key component in the body's immune system – plummeted following exposure to the gas. Rhoads wondered whether mustard gas, or other sulphur mustards, might be useful in treating leukaemia or lymphomas – types of cancer characterized by the production of excessive numbers of white blood cells. Shortly thereafter, a number of patients suffering from lymphomas back

The German bombing of Allied shipping in Bari in December 1943 had an unexpected consequence: the development of the first anti-cancer drugs.

in America were injected with sulphur mustards, and showed remarkable, albeit temporary, improvements in their condition. These were the first trials in treating cancers by chemotherapy, rather than surgery.

1944

Physics

AN UNCERTAIN OUTCOME

Werner Heisenberg, the German physicist who was involved in the Nazi nuclear energy programme, visited Zurich in neutral Switzerland to deliver a lecture. The American Office of Strategic Services (precursor of the CIA) dispatched one of its agents, a former Major League baseball player called Moe Berg, to attend the lecture. Berg had only ever been an average player, but he was a graduate of Princeton and Columbia Law School, read ten newspapers a day and spoke several languages, earning him the reputation as 'the brainiest guy in baseball'. In 1943 he joined the OSS and was tasked with capturing and interviewing Italian physicists to see what they knew about the Nazi nuclear programme, and about Heisenberg in particular. When news filtered through that Heisenberg was going to be in Zurich, Berg was told to go to the lecture and determine 'if anything Heisenberg said convinced him the Germans were close to a bomb'. If he concluded that this was the case, he was to shoot Heisenberg on the spot. Berg was not convinced, and Heisenberg survived the war to continue his glittering career until his death in 1976. As for Berg, after the war he never held down a proper job, and instead lived off relatives and friends. When asked what he did for a living, he would put his finger to his lips. He had rejected the award of the Medal of Freedom in 1945, but after his death in 1972 his sister accepted it on his behalf.

 **THE COST OF SPACE FLIGHT**

(7 September) The first V-2 rocket – Hitler's 'vengeance weapon' – was launched against England. The V-2, designed by Wernher von Braun, a member of the Nazi Party and the SS, was the first craft to make a suborbital spaceflight. In all, over 3000 V-2s were launched against Allied targets, causing the deaths of some 7250 people. However, many more slave labourers – some 20,000 – died in the manufacture of the rockets at the Mittelbau-Dora plant-cum-concentration camp. Some were shot or hanged, while many died of exhaustion, starvation or disease.

After the war, von Braun was whisked away to the United States, and became a naturalized US citizen – and the principal engineering genius behind the giant Saturn V booster rocket that sent the first men to the Moon.

BANDIT AT NINE O'CLOCK

(July) Such was the tension and paranoia among the members of the Manhattan Project prior to the first atomic test that one morning, when a bright object appeared low above the horizon, virtually everybody ran outside to look. The US Air Force base at Kirtland Field in Albuquerque was alerted, but reported that they had no interceptors that could come within range of the object. At this point, Robert Oppenheimer, director of the project, picked up the story in a letter to Eleanor Roosevelt: 'Our director of personnel was an astronomer and a man of some human wisdom; and he finally came to my office and asked whether we would stop trying to shoot down Venus.'

OPPENHEIMER REPENTANT

(16 July) When Robert Oppenheimer (*see above*) witnessed the first atomic test, he quoted a line from the *Bhagavad Gita*: 'Now I am become Death, the destroyer of worlds.' After the bombs were dropped on Hiroshima and Nagasaki, Oppenheimer was invited to meet President Truman in the Oval Office. Oppenheimer wrung his hands and exclaimed, 'I have blood on my hands.' Afterwards, Truman turned to an aid and exploded: 'Never bring that f***ing cretin in here again. He didn't drop the bomb. I did. That kind of weepiness makes me sick.' Oppenheimer went on to lobby for international control of nuclear power, which led to him being regarded as 'soft on the Reds', and in 1954 his security clearance was revoked.

SAFE CRACKING, PLATE JUGGLING AND QUANTUM ELECTRODYNAMICS

During the Second World War the young physicist Richard Feynman was employed in a non-central role on the Manhattan Project, which developed the atomic bomb. He found life in the isolation of Los Alamos boring, so entertained himself by learning to pick locks and crack safes. This last he found surprisingly easy, once he realized that physicists tend to adopt important numbers for their combinations, such as 27-18-28, derived from the base of natural logarithms, $e = 2.71828 \ldots$ He would alarm his security-conscious colleagues by leaving little notes in their safes alongside the papers detailing atomic secrets.

After the defeat of Japan, Feynman took up a post at Cornell University. Depressed after the dropping of the bombs on Hiroshima and Nagasaki, Feynman felt he had lost his sense of the fun of physics. Then one day his attention was caught when someone in the cafeteria threw a plate in the air. As it wobbled and span, Feynman noticed the spinning was faster than the wobbling, and that when the angle was very slight, the plate span precisely twice as fast as it wobbled. Working out the dynamics involved a complicated equation, and thus began Feynman's groundbreaking work that eventually led to the theory of quantum electrodynamics, a branch of physics that explains the interaction of light, in the form of photons, and matter, in the form of electrons. For this Feynman was awarded the 1965 Nobel prize for physics.

THE ALPHER–BETHE–GAMOW THEORY

1948
Cosmogony

With his student Ralph Alpher, the Russian-born US cosmologist and theoretical physicist George Gamow published 'The Origin of Chemical Elements' in the 1 April edition of *Physical Review*. Gamow also added the name of the German-born US physicist Hans Bethe to the paper (apparently without his knowledge or permission), to create a pun on the first three letters of the Greek alphabet, alpha, beta and gamma.

ON THE POSSIBILITY OF TIME TRAVEL – PART I

1949
Physics

The Austrian-American mathematician Kurt Gödel, building on Einstein's field equations in general relativity, showed that if the universe were rotating, conditions would be created whereby time looped back on itself, and so time travel would be possible. The flaw in this theory is that the universe does not appear to be rotating.

THE PEOPLE'S WAR AGAINST THE SNAIL

1949
Public health

The communist government of the newly established People's Republic of China launched 'The People's War Against the Snail'. The target was the freshwater snail living in paddy fields and waterways that harbours the parasitic worm causing the debilitating disease of schistosomiasis in humans. It was said that the peasants collected the snails one by one, using chopsticks. The campaign was the beginning of the successful elimination of the disease from many parts of China.

A Chinese government poster from 1954 advising the peasantry on ways to avoid contracting schistosomiasis – a parasitic infection carried by water snails.

1950 to 1989

The Bare-Fronted Hoodwink * On the importance of daydreaming * Demikhov's monster * A nuclear reactor in every home * On the irresistibility of sleep * Bombs for peace * Elephant on acid * On the efficacy of horseshoes * Echo of the big bang mistaken for pigeon poo * The Zambian mission to Mars * What turns turkeys on * Messages from the little green men? * On the insufferability of cosmologists * Heaven hotter than hell * Cello scrotum * The Wow! signal * Of God helmets and ghosts * The doctor who infected himself * Amoeba not so simple * A cure for hiccups

THE BARE-FRONTED HOODWINK

1950
Ornithology

Maury F.A. Meiklejohn published a scientific paper in *Bird Notes* describing this hitherto unknown bird species. The Bare-Fronted Hoodwink (*Dissimulatrix spuria*) has the ability to be 'almost seen', Meiklejohn noted, and can be identified by its 'blurred appearance and extremely rapid flight away from the observer'. On 1 April 1975 the Royal Scottish Museum in Edinburgh put a specimen of the Hoodwink on display. On 2 April it was revealed that its head was that of a carrion crow, its body that of a plover, and its feet those of an unidentified waterfowl.

ON THE IMPORTANCE OF DAYDREAMING

1953
Neuroscience

At the University of Pennsylvania the American neuroscientist Louis Sokoloff conducted an experiment in which, over the course of an hour, he monitored the brain activity and levels of oxygen and carbon dioxide in the jugular vein of a 20-year-old student. Sokoloff was surprised to find that the student's brain used as much oxygen while he was resting with his eyes closed as when he was carrying out difficult tests in mental arithmetic. Although the brain represents only 2 per cent of the body's mass, it uses 20 per cent of the calories we consume – whether we are concentrating hard on some difficult mental task, or just gazing vacantly into space. Further research in the 1980s and 1990s indicated that there were some areas of the brain that are extremely active when we are at rest, but which go quiet when we are faced with an immediate mental challenge. Some neuroscientists have concluded that the activity that shuts off is, in essence, daydreaming – part of which may involve sifting items for retention or otherwise in the memory.

In support of this, it has been found that when active, these parts of the brain use a disproportionate amount of glucose, perhaps as a building block for the creation and maintenance of synapses, the key connections that make up the memory.

FASTER THAN A PRAYER WHEEL

The JOHNNIAC computer – then the world's fastest – began operation. It was built by the Rand Corporation and named after John von Neumann, the man on whose architecture it was based: the acronym stands for *John von Neumann Numerical Integrator and Automatic Computer*. Neumann himself modestly declared: 'I don't know how really useful this will be. But at any rate it will be possible to get a lot of credit in Tibet by coding "*Om Mane Padme Hum*" a hundred million times in an hour. It will far exceed anything prayer wheels can do.'

MIND–CONTROL DRUGS

CIA director Allen Dulles set up Project MK-ULTRA to investigate the possibilities of using various drugs to control the minds of captives, or even foreign leaders. Experiments involving drugs such as LSD were often conducted without the subjects' knowledge or consent, and guinea pigs included CIA agents, military personnel and other government employees, mental patients, prostitutes and members of the general public. In one case, the CIA set up its own brothels and administered LSD to the unwitting clients, filming the results through one-way mirrors. The men involved, they calculated, would be unlikely to go public about their experiences. As it turned out, the effects of LSD proved to be far too unpredictable to make it an effective truth drug, and in many instances it only strengthened the subject's conviction that he or she could resist any form of interrogation, including torture. Other experiments involved heroin, marijuana, alcohol, sodium pentothal, barbiturates and amphetamines. In the case of the last two, one technique involved injecting a dose of barbiturate, and then when the subject was about to fall asleep, administering a dose of amphetamine. Just prior to the subject lapsing into total incoherence, it was sometimes possible to extract some helpful information. In 1973 CIA director Richard Helms ordered that all files relating to MK-ULTRA were to be destroyed, so the subsequent Congressional inquiry had very little to go on, beyond a handful of documents and the testimony of some of those involved.

DEMIKHOV'S MONSTER

Soviet surgeon Vladimir Demikhov grafted the head of a puppy onto the shoulder of an adult German shepherd dog. The German shepherd tried to shake off the puppy, which in turn bit back at its 'host'. When the puppy tried to lap milk, the milk oozed out from the base of its neck. The horrific creation survived for six days. Over the following years Demikhov repeated the experiment 19 more times. The techniques developed during the course of these experiments have been credited with pioneering the way for human transplant surgery.

HITTING A BRICK WALL

US test pilot Lieutenant Colonel John Stapp reached a speed of 1011 kph (632 mph) in a rocket sled, and then, when the braking system was applied,

In one of the most grotesque experiments ever undertaken, Vladimir Demikhov grafted the heads of puppies onto the shoulders of adult dogs.

for 1.1 seconds was subjected to a deceleration force of 25 G – a jolt greater than that experienced by a driver who hits a brick wall at 192 kph (120 mph). Stapp was unharmed, apart from the rupture of the capillaries in his eyeballs, but by the next day his eyesight had more or less returned to normal. He was secured from greater harm by the safety harness he was wearing, and for the rest of his life he campaigned for the compulsory fitting of seatbelts in automobiles.

1955
Post-mortem studies **EINSTEIN CARVED UP**

(18 April) Death of Albert Einstein. Thomas Stoltz Harvey, the pathologist at Princeton Hospital, where Einstein died, removed his brain for preservation,

believing that the neuroscientists of the future might be able to establish the physical basis of the great scientist's genius. What is less well known is that Einstein's ophthalmologist, Henry Abrams, removed Einstein's eyes, which are still preserved in a small glass jar in a safe deposit box in New Jersey. 'When you look into his eyes,' Abrams later told Einstein's biographer, Denis Brian, 'you're looking into the beauties and mysteries of the world.'

A NUCLEAR REACTOR IN EVERY HOME

(1 June) Robert E. Ferry, general manager of the American Institute of Boiler and Radiator Manufacturers, predicted that every home would soon have its own nuclear reactor. 'The system would heat and cool a home,' he said, 'provide unlimited hot water, and melt the snow from sidewalks and driveways.' All this, he said, could be provided over a period of six years at the cost of just $300 for the fissionable material. That same year, Alex Lewyt, a vacuum-cleaner manufacturer, predicted that within ten years vacuum cleaners would run on nuclear energy.

FIERY EXPECTORATION

(11 August) Death of the American physicist Robert W. Wood, known as the father of infrared and ultraviolet photography. A well-known practical joker, one of his favourite tricks was to spit into puddles and at the same time covertly drop in a fragment of sodium, which would react violently with the water – startling passers-by.

FAMOUS LAST WORDS

(September) The British Astronomer Royal, Sir Harold Spencer Jones, declared, 'Space travel is bunk.' Just two weeks later, on 4 October 1957, the Soviets launched *Sputnik 1*, Earth's first artificial satellite. An editorial in the following week's edition of *New Scientist* warned its readers not to get over-excited: 'It is very likely,' it opined, 'that generations will pass before man ever lands on the Moon.'

FOOD FOR THOUGHT

When the physicists C.N. (Frank) Yang of the Institute of Advanced Study at Princeton and Tsung-Dao Lee of Columbia University jointly won the Nobel prize for their work on the violation of parity law in weak interaction, the Chinese restaurant where they ate together every week put up a sign reading 'Eat here, get Nobel prize!'

YOU WHO ARE ABOUT TO DIE . . .

The US military developed a talking bomb. At 1200 m (4000 ft) above the target, parachutes would open and a tape recorder would switch on, booming out a propaganda lecture to those below who were about to be obliterated.

THE PAULI EFFECT

(15 December) Death of the Austrian theoretical physicist Wolfgang Pauli. Pauli, a pioneer of quantum physics, had given his name to the Pauli exclusion principle, and also, less seriously, to the Pauli effect. This suggests that in the presence of certain people, experimental equipment has a habit of undergoing catastrophic failure. So renowned was Pauli in this regard that his friend, the experimental physicist Otto Stern, banned him from his laboratory. Pauli himself was fascinated by the effect, and noted that when he was at Princeton in February 1950 the cyclotron caught fire.

But Pauli was clear what was, and what was not, science, on the grounds of provability; thus he once condemned a colleague's paper by saying, 'Not only is it not right, it's not even wrong.' When a less-esteemed physicist once interrupted him with a pedantic objection, Pauli put him down by saying 'Whatever you know I know.' When he first met Paul Ehrenfest, a more distinguished physicist, the latter remarked, after some talk, 'I think I like your papers better than you.' To which Pauli retorted, 'I think I like you better than your papers.' The two became good friends.

As he lay dying – at the age of only 58 – of a hitherto undiagnosed cancer, Pauli was disturbed that the number of his hospital room was 137. This number is of some significance in quantum physics, $1/137$ being the fine structure constant of the hydrogen spectrum.

ON THE IRRESISTIBILITY OF SLEEP

Sleep researcher Ian Oswald of Edinburgh University conducted an experiment on three young male volunteers. He taped their eyes open and placed a bank of flashing lights directly in front of their faces. At the same time, they were subjected to violent electrical shocks and extremely loud music. Nevertheless, once they were tired enough, even these extreme stimuli failed to keep them awake – or, as Oswald put it, 'There was a considerable fall of cerebral vigilance . . .'

IS THERE ANYBODY OUT THERE?

Dr Frank Drake, now professor emeritus of astronomy and astrophysics at the University of California, Santa Cruz, announced his eponymous equation at the meeting at Green Bank, Virginia, that launched SETI, the search for extraterrestrial intelligence. His equation aims to calculate how many extraterrestrial civilizations might be scattered across our own Galaxy, the Milky Way. He came up with the following formula to work out the number of civilizations (N):

$$N = N^* \times f_p \times n_e \times f_l \times f_i \times f_c \times f_L$$

where

N^* is the number of stars in the Galaxy;

f_p is the fraction of stars that have planetary systems;

n_e is the number of Earth-like planets (i.e. those that could potentially support life) in every star system;

f_l is the fraction of suitable planets on which life actually arises;

f_i is the fraction of life-bearing planets on which intelligent beings evolve;

f_c is the fraction of planets with intelligent beings who actually try to communicate across space;

f_L is the fraction of the planet's life for which the civilization survives.

The Drake equation does not provide the answer, but outlines a number of unknown factors that would need to be resolved before we could come up with any accurate estimate of the likelihood of hearing from extraterrestrial life forms. Given the very large number of stars in the Galaxy (100–400 billion), and the length of time the universe has been in existence (13.5–14 billion years) it would seem that the chances are relatively high, but this gives rise to the question, that if technologically advanced civilizations are that common in this or the 100 billion or so other galaxies, some with up to a trillion stars each, why haven't we heard from them? This is the so-called Fermi paradox, outlined by Enrico Fermi in 1950. The implication is that technologically advanced civilizations have a tendency to destroy themselves rather rapidly.

ESTABLISHING ONE'S CREDENTIALS

President Kennedy appointed the Nobel laureate Glenn Seaborg as chairman of the United States Atomic Energy Commission. At a subsequent congressional hearing, one hostile senator demanded of Seaborg, 'What do you know about plutonium?' To which Seaborg could reply, with complete truthfulness, that he had discovered it.

Seaborg was also involved in the discovery of a number of other transuranic elements: americium, curium, berkelium, californium, einsteinium, fermium, mendelevium and nobelium. The naming of element 106 – seaborgium – in his honour was for a while controversial, as some believed it was not right to name new elements after people still living. Seaborg himself commented on the controversy in 1995:

> There has been some reluctance on the part of the Commission for Nomenclature of Inorganic Chemistry of the International Union of Pure and Applied Chemistry to accept the name because I'm still alive – and they can prove it, they say.

Glenn Seaborg in his laboratory in 1980, showing the cigar boxes in which the very first samples of plutonium-239 were stored.

However, in 1997 the 39th IUPAC General Assembly in Geneva agreed to the name. Seaborg died two years later.

JUST OBEYING ORDERS – PART II

Stanley Milgram of Yale University, then a 27-year-old assistant professor, embarked on a highly controversial experiment, in which he wished to study whether people with an ordinary sense of moral decency could be suborned into carrying out atrocities. To this end he set up a scenario in which each subject believed they were participating in a study of memory and learning. The subject was told to put a series of questions to a 'learner' (in fact an actor), and if the 'learner' failed to provide the correct answer, the 'teacher' (the subject) was to give him what appeared to be an electric shock. As the experiment proceeded, the (non-existent) 'shocks' apparently became more and more severe. In many instances, as the 'learner' screamed and banged on the walls, the subjects turned to the 'experimenter' (the person supervising the proceedings) to say that they could not continue, but were calmly told to proceed with the experiment. When Milgram collated his results, he found to his astonishment that 65 per cent of his subjects proceeded up to the point where they delivered the maximum 'shock' – supposedly of 450 volts, way past the 375-volt level, where a warning sign read 'Danger: Severe Shock'. In a letter to his funding body, the National Science Foundation, Milgram wrote: 'I once wondered whether in all of the United States a vicious government could find enough moral imbeciles to meet the personnel requirements of a national system of death camps, of the sort that were maintained in Germany. I am now beginning to think that the full complement could be recruited in New Haven.'

BOMBS FOR PEACE

The USA dubbed its programme to develop peaceful uses for nuclear bombs 'Project Plowshare', after the well-known passage in the Old Testament: '. . . and they shall beat their swords into plowshares, and their spears into pruning-hooks' (Micah 4:3). A wide range of proposals was considered, including mineral extraction, linking underground aquifers, cutting through the Bristol Mountains of the Mojave Desert to carry Interstate 40 and a new railroad line, and blasting a new Pacific–Atlantic waterway through Nicaragua at sea level. In 1962 a proposal to use hydrogen bombs to create a massive artificial harbour on the north coast of Alaska was finally abandoned in the face of opposition by local Inuit people and the realization that such a harbour would have no use whatsoever. Under the auspices of Project Plowshare, a total of 27 test explosions were carried out in New Mexico, Colorado and Nevada between 1961 and 1973, including a number intended to access deposits of natural gas. However, it became clear that any gas extracted in this way would be too radioactive for safe use.

THE FUTURE OF COMMUTING

Air Chief Marshal Sir Ralph Cochrane, who had been in overall command of the Dambusters Raid in the Second World War, calculated that by 1976 the railways would be unable to cope with the numbers of people commuting to and from London. The solution, he proposed, comprised a fleet of 360 vertical take-off and landing (VTOL) aircraft, each carrying 110 passengers and flying at 960 kph (600 mph). He fully expected this system to be in operation by the end of the 20th century.

ELEPHANT ON ACID

(August) Warren Thomas, director of Lincoln Park Zoo in Oklahoma City, fired a dart into the rump of Tusko the elephant. The dart contained a dose of the hallucinogenic drug LSD 3000 times stronger than that used by humans for recreational purposes. The intention was to see whether LSD would induce the condition of 'musth', in which bull elephants enter a state of aggressive sexual frenzy, and exude a discharge from a gland between the ear and the eye. Things started well, with Tusko charging around and trumpeting loudly for a few minutes. Then he fell down dead. In their report to *Science* magazine, the experimenters concluded that 'It appears that the elephant is highly sensitive to the effects of LSD.'

ON THE EFFICACY OF HORSESHOES

1962
Physics

(18 November) Death of the Danish physicist Niels Bohr, the pioneer of quantum mechanics. When a colleague mocked him for hanging a good-luck horseshoe on the door of his country cottage, he retorted, 'Yes, it's nonsense, but they say it works even if you don't believe in it.'

CLARKE'S THREE LAWS

1962
Futurology

Arthur C. Clarke published *Profiles of the Future*, which included an essay entitled 'Hazards of Prophecy: The Failure of Imagination'. In this he proposed the First Law thus:

> When a distinguished but elderly scientist states that something is possible, he is almost certainly right. When he states that something is impossible, he is very probably wrong.

In the same essay, Clarke also observed:

> The only way of discovering the limits of the possible is to venture a little way past them into the impossible.

Others quickly dubbed this Clarke's Second Law. In the 1973 revised edition of *Profiles of the Future*, Clarke added a Third Law:

> Any sufficiently advanced technology is indistinguishable from magic.

A GALACTIC CIVILIZATION?

Nikolai Kardashev, the Soviet astrophysicist who was to become deputy director of the Russian Space Research Institute, suggested that CTA-102, an enormously powerful radio source discovered in the early 1960s by the California Institute of Technology, could be generated by a highly advanced extraterrestrial civilization. Such a civilization was defined as one that had not only found ways to utilize all the energy resources of its own planet (Type I civilization, according to Kardashev's classification), but also those of its own sun (Type II civilization), or even of its galaxy (Type III civilization). The radio source was subsequently identified as a quasar – a quasi-stellar radio source. Quasars are extremely energetic and extremely distant celestial objects, probably associated with the supermassive black holes at the centre of galaxies.

THE HALLUCINATIONS OF THE BLIND

University of Illinois medical scientists Alex E. Krill, Hubert J. Alpert and Adrian M. Ostfeld published the results of an experiment to test the effect of LSD on 24 totally blind people, aged between 16 and 76. They found that some of those who had not been blind from birth experienced visual hallucinations, including spots, flickers of light and colours (such as they occasionally experienced without the drug), or even more 'complex visual experiences', and concluded that a normal retina is not needed for the occurrence of LSD-induced visual hallucinations. None of those who *had* been blind from birth experienced any sort of visual hallucination. In all cases, the incidence of hallucinations affecting the other senses – smell, hearing, touch and taste – was much greater than in non-blind subjects.

One congenitally blind subject reported that the Braille he was reading appeared to be jumping off the page.

ECHO OF THE BIG BANG MISTAKEN FOR PIGEON POO

Arno Penzias and Robert Wilson, two physicists working at the Bell Laboratories in Holmdel, New Jersey, found a problem with their ultrasensitive microwave receiver, a giant horn antenna they were using for radio astronomy observations. The trouble was a constant interference, the same in all directions, and with a temperature of 2.7° Kelvin (i.e. just above absolute zero). At first they thought it might be caused by the proximity of New York City, and then they blamed the pigeons that roosted in the antenna, whose 'white dielectric material' (as Penzias described it) was encrusting the instrument. But even after the bird droppings were removed and the pigeons culled, the interference remained. Once the pair published their findings, other cosmologists realized that 2.7°K exactly matched the strength of the predicted cosmic microwave background radiation left over from the Big Bang. Wilson had been a student of Fred Hoyle, leading proponent of the steady-state theory of the universe, so had failed to make the connection – which was a crucial component in the general acceptance of the Big Bang theory that Hoyle so derided. Penzias and Wilson were awarded the Nobel prize for physics in 1978.

THE ZAMBIAN MISSION TO MARS

Around the time of his country's independence, a patriotic Zambian science teacher called Edward Makuka Nkoloso established the Zambia Academy of

Sciences and Space Technology. His aim was to send a Zambian expedition to Mars (or the Moon). 'Our rocket crew is ready,' he told the press:

> Specially trained space girl Matha Mwambwa, two cats (also specially trained) and a missionary will be launched in our first rocket. But I have warned the missionary he must not force Christianity on the people in Mars if they do not want it. One other difficulty has been holding us up. UNESCO has not replied to our request for £7,000,000, and we need that money for our rocket programme. Then we can lead world science.

While waiting for the funding to come through, training of his team continued:

> I'm getting them acclimatized to space travel by placing them in my space capsule every day. It's a 40-gallon oil drum in which they sit, and then I roll them down a hill. This gives them the feeling of rushing through space. I also make them swing from the end of a long rope. When they reach the highest point, I cut the rope – this produces the feeling of free fall.

The authorities declined Nkoloso's request to launch his rocket from Independence Stadium on Independence Day:

> If I had had my way Zambia would have been born with the blast of the academy's rocket being launched into space. But the Independence Celebrations Committee said that would terrify the guests and possibly the whole population. I think they were worried about the dust and noise.

The project seems to have fizzled out through lack of team motivation. 'I've had trouble with my space-men and space-women,' Nkoloso told the press in November 1964. 'They won't concentrate on space-flight; there's too much love-making . . .' It was subsequently reported that 17-year-old Matha Mwamba had become pregnant, and had been removed from the project by her parents.

KEEPING THINGS IN PERSPECTIVE

When Dorothy Hodgkin was awarded the Nobel prize for chemistry in recognition of her pioneering work in X-ray crystallography – particularly in revealing the structure of penicillin and vitamin B12 – the *Daily Mail* greeted the news with the immortal headline, 'Nobel Prize for British Wife'.

WHAT TURNS TURKEYS ON

In their book *Sex and Behaviour*, Martin Schein and Edgar Hale of Pennsylvania State University described the experiments they had conducted to find out what it was about female turkeys that turned on male turkeys. They found that males would mate enthusiastically with a lifelike model of a female. Then, to find out which aspect of the female form was responsible for triggering their amorous responses, the two scientists gradually removed bits of the model until it was reduced to just a head on a stick. Still the males would enthusiastically attempt copulation. Schein and Hale concluded that when a male turkey mounts a female, all he can see of her under his own larger body is her head, and that the sight of this provides sufficient stimulus to keep things moving along.

MESSAGES FROM THE LITTLE GREEN MEN?

Much to her astonishment, postgraduate research student Jocelyn Bell Burnell discovered that a point in the constellation of Vulpecula was emitting bursts of radiation precisely every 1.3373 seconds. The apparently unnatural regularity of the pulses led Bell and her PhD supervisor, Professor Antony Hewish, to dub the source LGM-1 – short for 'Little Green Men 1', jokingly speculating that the pulses were signals from some extraterrestrial civilization. The pulses, which swept across the Earth like light from a lighthouse, in fact turned out to be beams of electromagnetic radiation from a rapidly rotating, highly magnetized neutron star. Neutron stars of this type were subsequently named 'pulsars', short for 'pulsating stars'. In 1974 Professor Hewish was awarded the Nobel Prize, but Bell was not – despite the fact that it was she who made the initial discovery.

A DIFFERENT MIND

(2 November) Death of the New Zealand mathematician A.C. Aitken, who had been professor at Edinburgh University. He had a remarkable memory – he could recite the first 1000 digits of pi and had memorized the entire *Aeneid* while at school. He was also a prodigious mental calculator, and it was said of him, as it was of Ramunujan (*see* 1920), that he claimed the integers as personal friends. On one occasion he was challenged to express $^4/_{47}$ as a decimal. After a four-second pause, he began to spew out digits at a rate of one every three-quarters of a second: 0.0851063829787234042553191 4. He then paused to talk about the problem for a minute, then resumed: 191489. He paused again, just for five seconds, then continued: 361702127659574468. He then declared that at that point the sequence of digits started to repeat itself.

Aitken was also a passionate advocate of the duodecimal system, which he believed should sweep aside the decimal system. In *The Case Against Decimalization* (1962) he argued:

> But the final quantitative advantage, in my own experience, is this: in varied and extensive calculations of an ordinary and not unduly complicated

kind, carried out over many years, I come to the conclusion that the efficiency of the decimal system might be rated at about 65 or less, if we assign 100 to the duodecimal.

Quite what Aitken rated as an 'ordinary and not unduly complicated' calculation one can only guess at. Other advocates of the duodecimal system have included George Bernard Shaw, H.G. Wells and Sir Isaac Pitman.

INSIGNIFICANCE

Martin Gardner published *The Numerology of Dr Matrix*, a collection of his mathematical games columns in *Scientific American*. In one of these he points out that if you take the Dewey Decimal classification for numerology, 133.335, reverse it to get 533.331, then add the two numbers together, you get 666.666 – the Number of the Beast, twice. Gardner also pointed out that the Dewey Decimal classification for number theory was 512.81, whose two parts can be written as 2^9 and 9^2. The classification has since changed to 512.73.

ON THE INSUFFERABILITY OF COSMOLOGISTS

(1 April) Death of the Soviet physicist Lev Landau, known for his work in quantum mechanics, superfluidity and superconductivity. He famously observed that 'Cosmologists are often in error, but never in doubt.'

ON THE POSSIBILITY OF TIME TRAVEL – PART II

The American mathematical physicist and cosmologist Frank Tipler demonstrated that if an extremely massive and infinitely long cylinder spins at an immense speed, then you have the making of a time machine, for if you were then to fly in a spaceship on a spiral path around the cylinder, you could travel forward or backward in time, depending on the direction of the spiral. Tipler conceded that building such a machine was considerably beyond our current capabilities.

HEAVEN HOTTER THAN HELL

An anonymous article appeared in *Applied Optics* addressing the vexed question as to whether Heaven or Hell is the hotter destination. The author based his findings on two Biblical passages. The first, from Isaiah 30:26, reads: 'Moreover, the light of the moon shall be as the light of the sun and the light of the sun shall be sevenfold as the light of seven days.' From this the author concludes that 'Heaven receives from the Moon as much radiation as the Earth does from the Sun, and in addition 7 times 7 (49) times as much as the Earth does from the Sun, or 50 times in all'. Using the Stefan-Boltzmann fourth-power law for radiation, he calculated the temperature of Heaven to be 525°C. The second passage is from Revelations 21:8, which describes the damned condemned to a 'lake which burneth with fire and brimstone'. The author concludes from this: 'A lake of molten brimstone [sulphur] means that its temperature must be at or below the boiling point, which is 444.6°C. (Above that point, it would be a vapour, not a lake.)' Thus Heaven is hotter than Hell – or that at least was the consensus until 1998, when Jorge Mira Pérez and Jose Viña wrote to *Physics Today* to point out that Eugenio Ramiro Pose, Auxiliary Bishop of Madrid and Titular Bishop of Turuda, had clarified that only a single factor of 7 was intended in the Isaiah passage, not '7 times 7', and thus the temperature of Heaven is a more temperate 231.5°C – considerably cooler than Hell.

A NATURAL NUCLEAR REACTOR

1972
Nuclear physics

At the Pierreplatte uranium enrichment facility in France, tests showed that uranium from the Oklo mine in Gabon, Central Africa, contained a lower proportion of the fissionable U-235 isotope than normal. An investigation was immediately ordered, as any missing U-235 might have been used to make atomic weapons. Further examination of the uranium mined at Oklo compared to uranium from other mines showed that the former had a consistently lower concentration of U-235. This reduction in U-235 is precisely what occurs after nuclear fission, and the scientists of the French atomic energy commission concluded that, some 2 billion years ago, over a period of a few hundred thousand years, a series of natural self-sustaining nuclear chain reactions had taken place in a layer of uranium ore at Oklo. At the time, U-235 made up some 3 per cent of natural uranium, making a chain reaction possible; however, radioactive decay over the intervening period has reduced the concentration of U-235 to around 0.7 per cent – too low a concentration for natural nuclear reactions to take place.

CELLO SCROTUM

The *British Medical Journal* published the following letter:

> Sir,
>
> Though I have not come across 'guitar nipple' as reported by Dr P. Curtis (27 April, p. 226), I did once come across a case of 'cello scrotum' caused by irritation from the body of the cello. The patient in question was a professional musician and played in rehearsal, practice or concert for several hours each day. I am, etc.,
>
> J.M. Murphy
>
> Chalford, Glos.

Over the years, 'cello scrotum' was referenced in a number of other medical journals, and in one instance the writer questioned whether it might not be contact with the chair, rather than the instrument, that caused the condition. In 2009 Dr Elaine (now Baroness) Murphy admitted that she had written the letter, submitting it under the name of her husband, and that it was a hoax. 'Anyone who has ever watched a cello being played,' she said, 'would realize the physical impossibility of our claim.' A spokesman for the *BMJ* said, 'We may have to organize a formal retraction or correction now. Once these things get into the scientific literature, they stay there for good. But it all adds to the gaiety of life.'

SWEET SERENDIPITY

The artificial sweetener sucralose was discovered by chance by Shashikant Phadnis, a research student at Queen Elizabeth College (now part of King's College), London. Phadnis's supervisor, Professor Leslie Hough, had asked him to test the synthetic derivative of sucrose that they had made. Phadnis, whose first language was not English, thought Hough had asked him to 'taste' it – and so discovered a substance 600 times sweeter than sucrose.

1977
Astronomy

THE WOW! SIGNAL

(15 August) Dr Jerry R. Ehman, examining computer printouts from the Big Ear radio telescope of Ohio State University, detected a strong narrowband radio signal coming from the constellation Sagittarius, near the Chi Sagittarii star group. It lasted 72 seconds, and is as close to an artificial (i.e. intelligently generated) signal that has ever been detected. Ehman wrote 'Wow!' next to the signal on the printout, and this has become the name of the signal. However, it has never been heard again. Ehman himself says he is reluctant to draw 'vast conclusions from half-vast data'.

Dr Ehman's famous annotation on the printout containing the best candidate so far for an intelligent signal from outer space.

SAFER THAN EATING, OR SLEEPING WITH YOUR WIFE

With impeccable logic the Democratic governor of Washington, Dixy Lee Ray – who had been chairperson of the US Atomic Energy Commission in 1973–4 – stated:

> A nuclear power plant is infinitely safer than eating, because 300 people choke to death on food every year.

Two years later, Edward Teller – known as 'the father of the hydrogen bomb' – testified before an inquiry into the construction of Dresden III, a planned nuclear power plant in Illinois. He too was keen to emphasize the safety of nuclear power compared to everyday activities:

> What do you think you get more radiation from, leaning up against an atomic reactor or your wife? I don't want to alarm you, but all human beings have radioactive potassium in their blood – and that includes your wife . . . I do not advocate a law forcing couples to sleep in twin beds [but] from the point of view of radiation safety, I must advise you against the practice of sleeping every night with two girls, because then you would get more radiation than from Dresden III.

That same year Teller suffered a heart attack, which he blamed on the actress Jane Fonda and the anti-nuclear stance of her film *The China Syndrome*, which depicts a nuclear accident. The film came out just a week before the real nuclear accident at Three-Mile Island. Teller was unabashed, taking out a double-page newspaper advertisement in which he claimed: 'I was the only one whose health was affected by that reactor near Harrisburg. No, that would be wrong. It was not the reactor. It was Jane Fonda. Reactors are not dangerous.'

HOW BIG IS BIG?

The American mathematician Ronald Graham, while considering a problem in Ramsay theory, came up with a 'large number' as an upper bound for its solution. This number, now known as Graham's number, is too large to be expressed in conventional mathematical notation. Indeed, it is the largest number ever used in a mathematical proof, and the observable universe is much too small to contain a digital representation of it.

STARVED TO DEATH

(14 January) Death of the Austrian-American mathematician and logician Kurt Gödel, whose 'incompleteness theorem' showed that mathematics could never be a complete and self-consistent system, and who worked out the circumstances in which time travel would be possible (*see* 1949). Towards the end of his life, Gödel became increasingly paranoid, and would not eat anything that had not been tasted by his wife Adele, for fear that his food had been poisoned. Late in 1977 Adele became ill and had to be hospitalized. Gödel stopped eating altogether, and when he died he weighed only 29.5 kg (65 pounds). The cause of death was certified as 'malnutrition and inanition caused by personality disturbance'.

ODDEST TITLE OF THE YEAR

Establishment of the Diagram Prize for Oddest Title of the Year, a non-serious competition run annually by *The Bookseller* magazine. The winner in 1978 was *Proceedings of the Second International Conference on Nude Mice* (University of Tokyo Press). Scientific or technical books that have since won the prize include:

- *The Theory of Lengthwise Rolling* (1983, Mir Publishers);
- *Highlights in the History of Concrete* (1994, British Cement Association);
- *Developments in Dairy Cow Breeding: New Opportunities to Widen the Use of Straw* (1998, Nuffield Farming Scholarship Trust);
- *Weeds in a Changing World* (1999, British Crop Protection Council); and
- *High Performance Stiffened Structures* (2000, Professional Engineering Publishing).

HOW MUCH CAN WE KNOW?

In his book *Broca's Brain: The Romance of Science*, Professor Carl Sagan calculated how many bits of information the human brain could carry. There are some 10^{11} neurons in the brain, he wrote, each with around 1000 dendrites, tiny 'wires' that connect each neuron with the network of other neurons. On the assumption that each of these connections represents one bit, then the brain can accommodate 10^{14} items of information – far from enough, Sagan points out, to list the three-dimensional positions of the 10^{16} atoms of chlorine and sodium in a microgram of table salt. Luckily, nature has laws that we can uncover, so to understand what is happening in a microgram of salt, we do not need to create a complete representation of it – otherwise, to have any understanding of the Universe and its 10^{80} particles, we would need a brain at least as big as the Universe.

A HORSE-POWERED
HORSELESS CARRIAGE

A British inventor called Philip Barnes filed a patent for a road vehicle that would be powered by a horse. The horse would not pull the vehicle, but rather be placed inside it, where it would walk along a conveyor belt that would turn the wheels.

SEXING THE BREATH

A study at the University of Pennsylvania demonstrated that in a series of blindfold tests 95 per cent of subjects could correctly tell the gender of the person breathing on them.

OF GOD HELMETS
AND GHOSTS

Michael Persinger, a neuroscientist at Laurentian University, Canada, proposed a hypothesis that mystical and religious experiences are 'evoked by transient, electrical microseizures within deep structures of the temporal lobe'. Persinger designed experiments in which subjects wore a modified snowmobile helmet equipped with solenoids that generated weak but complex magnetic fields in the vicinity of the subject's temporal lobes – the areas of the brain involved with memory and with visual and auditory perception. Some 80 per cent of his subjects reported the presence of another being in the room, sometimes identified as a dead loved one, or even God.

Dr Susan Blackmore wearing a 'God helmet', in which complex magnetic fields induce what are perceived of as 'paranormal' experiences.

Two decades later, Jason Braithwaite, a cognitive psychologist at the University of Birmingham with an interest in people's apparent experiences of the paranormal, investigated the Tapestry Room at Muncaster Castle in Cumbria, where many overnight guests had reported strange experiences, such as hearing the cries or screams of children, or strange footsteps, or had felt themselves touched by an unknown presence. On the hunch that unusual magnetic fields might account for such experiences, Braithwaite set up two magnetometers that could detect even extremely weak magnetic fields. He indeed found some particularly complex magnetic fields in the Tapestry Room, and discovered that these were associated with the iron mesh in the bedstead underneath the mattress. As someone in the bed turned over, the resulting movement in the iron mesh caused wild fluctuations in the magnetic fields around the head of the bed, fluctuations that were comparable to those generated in Persinger's 'God helmet'. Perhaps it was this that had caused all those terrifying 'paranormal' experiences.

NOT JUST AN ANTIDEPRESSANT . . .

1981
Pharmacology

Three psychiatrists from the St John Regional Hospital, New Brunswick, reported in the *Canadian Journal of Psychiatry* an unexpected side effect experienced by a small proportion of patients taking the antidepressant clomipramine. Whenever they yawned, these patients said, they had an orgasm. One woman 'sheepishly admitted that she hoped to take the medication on a long-term basis', while one man was obliged to wear a condom permanently, in case he encountered a bore on the bus. A middle-aged woman who was hospitalized begged to be taken off the medication, as in the hospital environment there was no satisfactory way to deal with her 'unresistable [sic] sexual urges'. A fourth patient, a married man, reported a rather different experience on the drug: every time he yawned 'he experienced such an intense sense of exhaustion and weakness, that he had to lie down for 10 to 15 minutes'.

The following year, the *Canadian Journal of Psychiatry* published some comments by another team of scientists on the phenomenon, providing a possible explanation: 'We propose that the increase in brain serotonin levels resulting from clomipramine's effect on serotonin re-uptake may stimulate release of hypothalamic CRF [corticotrophin releasing factor]. The CRF released has the potential to activate neural circuits responsible for the previously described behaviour patterns, i.e. yawning and spontaneous sexual response.'

HOW COLD?

(20 October) Death of the brilliant British theoretical physicist Paul Dirac, who predicted the existence of antimatter and shared the 1933 Nobel prize with Erwin Schrödinger. Dirac's abilities were confined to the lofty plane of his work; socially, he was at a tangent to the rest of the world. At Cambridge, a classics don tried an opening conversational gambit by remarking 'Cold, isn't it?' Dirac paused, thought, and then responded, 'How cold?' In 1929, while sailing from America to Japan with Werner Heisenberg, he asked his companion why he danced every night. Heisenberg replied, 'Well, when there are nice girls, it is a pleasure to dance.' After some minutes of deep thought, Dirac asked, 'But how do you know beforehand that the girls are nice?' Dirac blamed his lack of social skills and awkward personality – he may have been autistic – on his bullying father. 'I never knew love or affection when I was a child,' he admitted.

THE DOCTOR WHO INFECTED HIMSELF

For years, physicians had believed that gastric and duodenal ulcers (collectively known as peptic ulcers) were caused by factors such as stress and bad diet. But Barry Marshall, medical registrar at the Royal Perth Hospital in Western Australia, and his colleague, the pathologist Robin Warren, had found that the bacterium *Helicobacter pylori* was present in all patients with duodenal ulcers, and in three-quarters of those with gastric ulcers. Their findings were dismissed by many medical scientists, who maintained that *H. pylori* could not be responsible for peptic ulcers, as no bacteria could survive in the acidic environment of the stomach. Having failed to infect piglets with the bacterium, Marshall decided to infect himself, and swallowed a Petri dish of *H. pylori*. After a few days he began to experience discomfort, nausea, vomiting and bad breath – the symptoms of gastritis, which often leads to ulcers. After ten days his wife insisted he take antibiotics, and the symptoms disappeared. It took some years for his findings to be generally accepted, but now peptic ulcers are routinely treated with antibiotics. In 2005 Marshall and Warren were awarded the Nobel prize.

A LASER SHAVER

(26 November) Eugene J. Politzer filed a patent for a device to burn off stubble using a laser. To quote United States Patent 4819669:

> A method of shaving a beard is disclosed which includes the steps of passing the hairs of a beard through an electrically and thermally insulating grid and then applying laser energy of selected wave lengths to the ends of the

275

hairs passing through the grid. Preferably, the energy from the laser source is directed perpendicularly to the hairs in the grid.

A patent was issued on 11 April 1989, but the idea has not been taken up by manufacturers.

AN APT NAME

While investigating the biology of the Cueva de Villa Luz in Mexico, Jim Pisarowicz came up with an appropriate name, now generally adopted, for the colonies of sulphur-synthesizing bacteria that hang like stalactites from the ceiling of this and certain other caves, dripping highly concentrated sulphuric acid onto the floor. He called them snottites.

AMOEBA NOT SO SIMPLE

In his book *The Blind Watchmaker*, Richard Dawkins points out that some species of amoeba have as much information in their DNA as 1000 volumes of *Encyclopaedia Britannica*.

A CURE FOR HICCUPS

The journal *Annals of Emergency Medicine* published a paper by Francis M. Fesmire MD of University Hospital, Jacksonville, Florida, entitled 'Termination of Intractable Hiccups with Digital Rectal Massage'.

STAR IN A JAR

(23 March) Stanley Pons and Martin Fleischmann gave a press conference at the University of Utah announcing their achievement of 'cold fusion' – the recreation at room temperature of nuclear fusion, the energy that powers the Sun, using just two electrodes and a test tube of special water. No one else was able to repeat their experiment, which would have broken the laws of thermodynamics, and which was considered by many the modern equivalent of the medieval alchemists' claims to be able to turn base metals into gold.

REGGIO CALABRIA SYNDROME

After three decades of conducting autopsies on dead Mafiosi in the southern Italian city of Reggio Calabria, Francesco Aragona of the University of Messina concluded that the internal organs of gangsters show levels of stress – such as enlargement and ulceration – comparable to those of 70-year-old victims of strokes or heart attacks. The press dubbed this 'Reggio Calabria syndrome', the city being infamous as the 'murder capital of Italy', but when in 1996 lexicographers at HarperCollins suggested that the term might be included in forthcoming editions of *Collins English Dictionary* there was outrage. 'This is a piece of pseudo-culture,' declared Reggio's deputy bishop, Salvatore Nunnari. 'It presumes to make judgments that have no relationship to reality.' As for Francesco Aragona, he was proud of the fact that he could tell Mafiosi from innocent victims just by examining their innards.

1990 to the future

Bucky balls baffle peers * The Friendly
Floatees * Blowing up one's meat * On
the dangers of dihydrogen monoxide
* US air force plans 'gay spray' * The
perfection of synthetic faeces * The
pantomime moose of Wyoming *
The Bloop * Arsole, fukalite, nonanal
and other unlikely compounds * The
mystery of Witch's Hole * Pleasure by
remote control * Robot's flesh-eating
tendencies denied * Fish FRTs * Lazy
rays defy Einstein * The Time Traveller
Convention * The world's first penis
transplant * A taste of raspberries at
the centre of the Milky Way

GALILEO LESS RATIONAL THAN HIS PERSECUTORS, SAYS FUTURE POPE

Cardinal Josef Ratzinger – the future Pope Benedict XVI – delivered a speech in which he said that Galileo had been dogmatic and sectarian in his assertions that the Earth revolved around the Sun, and that the Church's trial of him for heresy was both 'reasonable and just'. The cardinal quoted the Austrian philosopher Paul Feyerabend, who said, 'At the time of Galileo, the Church remained more faithful to reason than Galileo himself.' At his trial in 1633, the Inquisition concluded that Galileo's view of the solar system was 'absurd, philosophically false, and formally heretical, because it is expressly contrary to Holy Scriptures'. To avoid being burnt at the stake, Galileo was forced to recant, and spent the rest of his life under house arrest.

A MODEST MAN

(30 January) Death of the quietly spoken scientist John Bardeen, 'Whispering John', whose two passions were physics and golf. He kept the two separate, to the extent that one day his long-term golf partner asked him, 'Say, John, I've been meaning to ask you. Just what is it you do for a living?' Bardeen had omitted to mention that he had won not just one, but two Nobel prizes: the first in 1956 for the invention of the transistor, and the second in 1972 for explaining superconductivity.

BUCKY BALLS BAFFLE PEERS

During a debate in the House of Lords on whether the British government should provide more funding for research on buckminsterfullerene (C_{60}), the then newly discovered fourth allotrope of carbon, Baroness Sear asked whether the subject of the debate was 'animal, vegetable or mineral'. Lord Renton, vaguely aware of its structure, asked whether it was 'in the shape of a rugger football or a soccer football', while Lord Campbell of Alloway plaintively inquired, 'What does it do?' To which Earl Russell, quoting Gilbert and Sullivan's characterization of the House of Lords in *Iolanthe*, replied that it 'does nothing in particular and does it very well'. It was explained to their lordships that buckminsterfullerene took its name from the American architect, Buckminster Fuller, creator of the geodesic dome, to which the C_{60} molecule bears a resemblance.

THE IG NOBEL PRIZES

The Annals of Improbable Research founded the annual Ig Nobel prizes, a parody of the Nobel prizes, awarded to scientists and others who 'first make people laugh, and then make them think'. Award winners over the years have included:

- Alan Klilgerman, for inventing Beano, an enzyme-based dietary supplement that helps to prevent flatulence (medicine, 1991).
- A team from the Shiseido Research Centre of Yokohama, for concluding that those who think they have foot odour, do, while those who think they don't, don't (medicine, 1992).
- E. Topol and his 975 co-authors for 'An International Randomized Trial Comparing Four Thrombolytic Strategies for Acute Myocardial

Infarction' (*New England Journal of Medicine*), a paper that had 100 times as many authors as it did pages (literature, 1993).

- Paul Williams of the Oregon State Health Division and Kenneth W. Newel of the Liverpool School of Tropical Medicine for their paper 'Salmonella Excretion in Joy-Riding Pigs' (biology, 1993).
- The Southern Baptist Church, who presented a county-by-county estimate of how many citizens of Alabama would go to Hell if they did not repent (mathematics, 1994).
- A team from Keio University, Japan, who succeeded in training pigeons to tell a Picasso from a Monet (psychology, 1995).
- Anders Barheim and Hogne Sandvik of Bergen University for their paper 'Effect of Ale, Garlic and Soured Cream on the Appetite of Leeches' (biology, 1996).
- An international team based in Switzerland, Japan and the Czech Republic for their study of people's brainwave patterns as they chewed different flavours of gum (biology, 1997).
- Bernard Vonnegut of the State University of New York at Albany for his report, 'Chicken Plucking as Measure of Tornado Speed' (meteorology, 1997).
- Patient Y and his doctors at the Royal Gwent Hospital, Newport, for a report entitled 'A Man Who Pricked His Finger and Smelled Putrid for Five Years' (medicine, 1998).
- Dr Len Fisher of Bath and Sydney, who calculated the optimum way to dunk a biscuit, and Professor Jean-Marc Vanden-Broeck of the University of East Anglia, who calculated how to make a teapot spout that does not drip (physics, 1999).
- A team from the University of Pisa and the University of California, San Diego, who established that from a biochemical point of view, romantic love is indistinguishable from obsessive-compulsive disorder (chemistry, 2000).
- Chris Niswander of Tucson, Arizona, for his software that alerts you when your cat walks across your computer keyboard (computer science, 2000).

- The US televangelist Dr Jack Van Impe for his assertion that 'black holes fulfill all the technical requirements to be the location of Hell' (astrophysics, 2001).
- Arnd Leike of the Ludwig Maximilian University of Munich, who showed that the froth on top of a glass of beer obeys the mathematical law of exponential decay (physics, 2002).
- Yukio Hirose of Kanazawe University for his chemical analysis of a bronze statue in the city that remained mysteriously unvisited by pigeons (chemistry, 2003).
- Victor Benno Meyer-Rochow of the International University Bremen, Germany, and Jozsef Gal of Lorand Eotvos University, Hungary, for their paper entitled 'Pressures Produced When Penguins Poo – Calculations on Avian Defecation', which was inspired by the ability of penguins to bend over and eject their faeces some distance from their nests (fluid dynamics, 2005).
- Patricia V. Agostino, Santiago A. Plano and Diego A. Golombek, for showing that hamsters recover from jetlag more quickly if given Viagra (aviation, 2007).
- Dan Ariely of Duke University, North Carolina, for showing that expensive placebos are more effective than cheap ones (medicine, 2008).
- Geoffrey Miller of the University of New Mexico, for showing that lap dancers are given bigger tips when they are ovulating (economics, 2008).
- Toshiyuki Nakagaki of Hokkaido University, for showing that slime mould can navigate a maze (cognitive neuroscience, 2008). He did not understand why his team had been given the award. 'We are always serious,' he said.
- Sharee Umpierre at the University of Puerto Rico for showing that Coca Cola is a spermicide, and Chuang-Ye Hong at Taipei Medical University for showing that it is not (chemistry, 2008).

THE FRIENDLY FLOATEES

Thirty thousand yellow plastic ducks intended as children's bath toys were liberated into the Pacific Ocean when their container was washed off the ship carrying them from China. Some of the so-called Friendly Floatees made their way through the Bering Strait into the Arctic Ocean, from where they eventually drifted south into the Atlantic. In 2007 some of them arrived off southwest England; others had reached Indonesia, Australia and South America. The ducks, now bleached white by the elements, can command prices of up to £500, such is their collectability. Their distribution and progress through the oceans has provided Seattle oceanographers Curtis Ebbesmeyer and James Ingraham with invaluable information about ocean currents.

BLOWING UP ONE'S MEAT

A team of US scientists led by Morse Solomon of the Department of Agriculture's Agricultural Research Service devised a new method of tenderizing tough meat. A chunk of particularly unchewable beef was placed on a steel plate at the bottom of a plastic drum of water, in which a small amount of explosive – roughly equal to a quarter of a stick of dynamite – was detonated. The steel plate reflected the resulting shock waves, shattering the tough fibres in the meat. The device, called 'the Hydrodyne', has the added benefit of destroying any harmful bacteria, and by 1998 comprised a 1244-litre (280-gallon) drum in which 273 kg (600 lb) of 'subprimal beef' could be tenderized in a single explosion. The technique has not yet found a place in domestic kitchens.

HAWKING BANS TIME TRAVEL

Stephen Hawking formulated his 'chronology protection conjecture', which proposes that the fundamental laws of physics prevent time travel. Creating loops in time via which time travel could occur, he suggested, would, via a kind of negative feedback, cause physical barriers within the loops. Thus you could not travel back in time and kill your grandmother before she had conceived your mother.

FLAMES IN SPACE

In order to study what happens to flames in space, scientists from the Fire Research Station in Garston, Hertfordshire, undertook a series of experiments under conditions of near-zero gravity aboard the 'vomit comet' –the nickname of various aircraft that generate nearly weightless conditions during parabolic flights. They found that gas flames 'became spherical like a bubble and turned from yellow to blue, finally becoming invisible'. The flames from small pieces of paper that had been set alight 'died back and went out'. The explanation for this is that in the absence of gravity-driven convection, oxygen cannot reach the fire unless driven by a fan.

A LIKELY STORY

In his book *The Unnatural Nature of Science*, Professor Lewis Wolpert cited a number of examples of the counter-intuitive results found in science. These results are sometimes related to the unimaginably large numbers involved.

In one example he takes the case of a glass of water, which is then emptied into the sea and its contents allowed to disperse through all the world's oceans. When the glass is refilled from the sea, it is almost certain that it will contain some of the original water molecules. 'What this means,' Wolpert concludes, 'is that there are many more molecules in a glass of water than there are glasses of water in the sea.'

1992
Chemistry /
public health

ON THE DANGERS OF DIHYDROGEN MONOXIDE

A new website appeared calling for a ban on DHMO – dihydrogen monoxide. The site warned that the chemical:

- is the major component of acid rain.
- contributes to the greenhouse effect.
- may cause severe burns.
- contributes to the erosion of our natural landscape.
- accelerates corrosion and rusting of many metals.
- may cause electrical failures and decreased effectiveness of automobile brakes.
- has been found in excised tumours of terminal cancer patients.

Despite the danger, the site continued, DHMO continues to be used:

- as an industrial solvent and coolant.
- in nuclear power plants.
- in the production of styrofoam.
- as a fire retardant.
- in many forms of cruel animal research.
- in the distribution of pesticides. Even after washing, produce remains contaminated by this chemical.
- as an additive in certain 'junk-foods' and other food products.

Dihydrogen monoxide is, of course, water (H_2O), and the website was a spoof created by Craig Jackson. Warnings about the dangers of DHMO had first been circulated in 1989 by Eric Lechner, Lars Norpchen and Matthew Kaufman, students at the University of California, Santa Cruz.

1993
Artificial intelligence

A CATHOLIC TURING TEST

Greg Garvey launched 'The Automatic Confession Machine: A Catholic Turing Test'. The Turing Test, devised by the computer pioneer Alan Turing in 1950, proposed that, to assess whether a machine had artificial

Greg Garvey's comment on the intrusion of technology into private life: an 'artificially intelligent' machine that substitutes for a priest in the confessional.

intelligence, a human should be placed in one room and the machine in another. If in communicating with the machine the human could not tell whether it was a machine or another human being, it could be said to possess artificial intelligence. In response to the Turing challenge, Garvey designed a machine that could replace the priest in the confessional. He describes it thus:

> This installation kiosk is a computerized confessional designed and fabricated to resemble an automatic banking machine. As with an ATM, the human computer interface (HCI) employs a simple alpha-numeric keypad and low resolution display. Human factors and religious ergonomics dictate the addition of a kneeler. A sinner's spiritual accounting requires selections from a menu of the seven deadly sins and the Ten Commandments. Forgiveness is computed and the user is [sic] receives appropriate penance as confirmation of the transaction.

Garvey is still developing the idea, and on his website says that 'Release 6.0.1 retains the look and feel of the original graphical user interface, written in Hypercard. Now updated with SuperCard 4.5.2 the ACM software is OSX compatible and will soon be deployed on most mobile devices.'

US AIR FORCE PLANS 'GAY SPRAY'

1994
Chemical warfare

The US Air Force's Wright Laboratory in Dayton, Ohio, submitted a number of proposals for the development of disabling chemicals to be sprayed on to enemy positions. 'One distasteful but non-lethal example,' the scientists at Wright suggested, 'would be strong aphrodisiacs, especially if the chemical also caused homosexual behaviour.' The idea – which carried a price tag of $7.5 million – was that the enemy would be so distracted that they would pay little attention to the job in hand.

THE PERFECTION OF SYNTHETIC FAECES

In order to test the company's range of nappies and incontinence pads, scientists at Kimberley-Clark in Dallas, Texas, sought to create an odourless and non-hazardous simulacrum of human faeces. Early experiments with pumpkin pie and peanut butter proved unsuccessful, as the solid and liquid ingredients separated too readily. After a period of intensive research, Richard Yeo and Debra Welchel came up with the ideal recipe: a mixture of starches, polyvinyls, gums, gelatin, resins and fibres. Just add water, and you get, according to Yeo, 'as close to the real thing as possible'.

NOT A BOMB

As France conducted nuclear tests in the Pacific, Jacques le Blanc, the French ambassador to New Zealand, sought to clarify matters. 'It is not a bomb,' he insisted. 'It is a device which is exploding.' The following year France signed the Comprehensive Test Ban Treaty.

OTTERS MISTAKEN FOR ENEMIES

Between 1981 and 1994, the Swedish navy's highly sensitive sonar detectors picked up 6000 incursions of Soviet and then Russian submarines into Swedish waters. But a government-appointed scientific commission reported in 1996 that just 1 per cent of these incidents could definitely be attributed to submarines. The remainder were almost certainly otters or mink paddling through the water, giving readings indistinguishable from those produced by revolving propellers.

THE PANTOMIME MOOSE OF WYOMING

In order to study how moose in Wyoming's Grand Teton National Park would react to the urine and faeces of predators – such as bears, wolves and big cats – that had long been absent from their territory, researcher Joel Berger and his wife and colleague Carol Cunningham from the University of Nevada found that there was only one way to get close enough to their subjects to find out. They commissioned Debra Markert, who had created various costumes for the Star Wars movies, to make them a moose suit. Draped with cameras, notepads and bags full of predator poop, Berger took the head end, and Cunningham the less enviable rear portion. Adopting the movements of a calmly feeding moose, and uttering the odd moose-like 'Mew', they found they could, on most occasions, approach their moose without arousing suspicion and deposit their smelly cargoes close enough for observation purposes. There was the odd exception. 'Carol and I were in deep snow,' recalls Berger, 'and a moose lowered its ears, dropped its head and the hair on its nape stood up. Think of a dog, when it gets nervous and its hackles stand up. That's basically what a moose does. And we were only about 15 yards away from it. And because we were in deep snow, it was like, uh oh. So we took the suit off. And the moose was very confused. Its demeanor changed. We went in opposite directions.' It turned out that most moose in Wyoming, unlike their cousins in Alaska who are still preyed upon by various carnivores, were undisturbed by the scents of their one-time enemies.

THE BLOOP

An array of hydrophones (underwater microphones) previously used by the US National Oceanic and Atmospheric Administration in the Cold War to track Soviet submarine activity picked up a mysterious sound. The extremely powerful ultra-low-frequency sound, originating off the southwest coast of South America and audible over distances of 5000 km (3100 miles), rose rapidly in pitch for a minute, then stopped. The 'Bloop', as it was dubbed, was detected on several occasions that summer, but has not been heard since. If it was made by an animal, it would need to be a creature much bigger than a whale, or considerably more efficient in producing sound. Also detected that year (on 19 May) was the 'Slow Down' sound, which dropped in frequency over a period of seven minutes. It too appears to have originated in the Pacific west of South America, and was audible some 2000 km (1240 miles) away.

ARSOLE, FUKALITE, NONANAL AND OTHER UNLIKELY COMPOUNDS

The University of Bristol set up a website to catalogue silly or unusual names given to chemical compounds, since then other websites have joined in this important work. Examples of entertaining chemical names include:

- Angelic acid (found in the plant angelica; there is also a diabolic acid);
- Arsole (similar to pyrrole, but with the nitrogen atom replaced by an arsenic atom);
- Cadaverine (a foul-smelling chemical produced during putrefaction);
- Complicatic acid (derived from the plant *Stereum complicatum*);

- Crapinon (an anticholinergic drug that can induce constipation as a side-effect);
- DAMN (the acronym of diaminomaleonitrile, which contains various cyanide groups);
- DEAD (diethyl azodicarboxylate, a compound that is explosive, carcinogenic and an irritant);
- Dickite (a mineral named after its discoverer, W. Thomas Dick);
- Draculin (an anticoagulant found in the saliva of vampire bats);
- Erectone (a compound used in Chinese medicine and extracted from the plant *Hypericum erectum*);
- FucK (L-fuculokinase, an enzyme);
- Fukalite (a mineral mined in the Fuka region of Japan);
- Gibberelin (a range of plant hormones involved in controlling growth);
- Jesterone (a chemical found in the fungus *Pestalotiopsis jesteri*);
- Khanneshite (a mineral found in Khanneshin, Afghanistan);
- Manxane (a compound resembling the three-legged emblem of the Isle of Man);
- Miraculin (the active constituent of the miracle fruit, which, if eaten, makes sour foods taste sweet);
- Moronic acid (3-oxoolean-18-en-28-oic acid, a natural triterpene);
- Nonanal ($C_9H_{18}O$, an aldehyde derived from nonane);
- Parisite (a mineral named after J. J. Paris);
- Penguinone (3,4,4,5-tetramethylcyclohexa-2,5-dienone, which in two-dimensional representations looks like a penguin);
- Performic acid (a strongly oxidizing acid related to formic acid);
- Psicose ($C_6H_{12}O_6$, a low-calorie sugar);
- Pterodactyladiene (a group of molecules resembling pterodactyls);
- Putrescine (another chemical produced during putrefaction);
- Rednose (a sugar derived from the degradation of the antibiotic rudolphomycin, which was named not after a reindeer, but, like a range of other antibiotics, after a character in Puccini's opera *La Bohème*);

- SEX (sodium ethyl xanthate, a flotation agent used in mining);
- Skatole (a noxious chemical found in scat, i.e. faeces);
- SnOT (the formula for tritiated tin hydroxide);
- Traumatic acid (a compound found in plants that helps to heal damaged tissue);
- Vomitoxin (a fungal toxin found in grains, although its main effect is to put animals off their feed).

DON'T DRINK AND DRIVE

Samsung developed voice-recognition technology for use in automobiles. Among its abilities, the system could detect whether the driver to whose voice it had become accustomed was drunk or sober. If even a trace of a slur was detected, the engine would refuse to fire, and instead a disembodied voice would deliver a short homily: 'Please don't drink and drive as you are putting your life and property at risk.'

ON THE UNRELIABILITY OF PARENTAL ADVICE

Arthritis and Rheumatism (Volume 41, issue 5) published a letter from Donald L. Unger MD of Thousand Oaks, California, in which Dr Unger outlined an important long-term experiment on himself:

> During the author's childhood, various renowned authorities (his mother, several aunts, and, later, his mother-in-law) informed him that cracking his knuckles would lead to arthritis of the fingers. To test the accuracy of this hypothesis, the following study was undertaken.

For 50 years, the author cracked the knuckles of his left hand at least twice a day, leaving those on the right as a control. Thus, the knuckles on the left were cracked at least 36,500 times, while those on the right cracked rarely and spontaneously. At the end of the 50 years, the hands were compared for the presence of arthritis.

There was no arthritis in either hand, and no apparent differences between the two hands . . .

This result calls into question whether other parental beliefs, e.g., the importance of eating spinach, are also flawed. Further investigation is likely warranted.

Dr Unger was anxious to point out that 'This study was done entirely at the author's expense, with no grants from any governmental or pharmaceutical source.'

THE NEW FINGERPRINTS

Research by the FBI concluded that every pair of jeans, once it has been worn a few times, acquires its own unique pattern of creases. This enabled them to identify masked robbers caught on CCTV, and led to at least one conviction.

BREASTFEEDING CONDEMNED AS 'INCESTUOUS'

At the Republican National Convention in the USA, a campaign was launched to ban breastfeeding, on the grounds that 'it is an incestuous relationship between mother and baby that manifests an oral addiction leading youngsters to smoke, drink and even becoming a homosexual'.

The campaign was sustained for two years, until American prankster and musician Alan Abel confessed that he had perpetrated it as a satire on American social conservatism.

THE MYSTERY OF WITCH'S HOLE

A survey of the Witch's Hole, a depression in the floor of the North Sea 150 km (90 miles) northeast of Aberdeen, found something rather surprising. The unmanned submarine beamed back pictures of an old trawler, sitting upright on the seabed, 150 m (80 fathoms) below the surface. There was no sign of damage, and it appeared to have gone down flat, rather than either end first – which would have been the case in the event of a collision or holing. The Witch's Hole is just one of many 'pockmarks' in the area. which is known as the Witch's Ground. These depressions in the seafloor sediments are caused by emissions of methane gas. It has been suggested that such an emission might have caused the sinking of the trawler, which was of a type built in the early 20th century. Methane is lighter than air, so it would be impossible for any boat to float on it. A big eruption of methane in the form of a giant bubble under a ship could thus cause it to drop like a stone down a lift shaft. Such a hypothesis, which has also been used to explain disappearances of ships within the so-called Bermuda Triangle, has not been substantiated by hard evidence, but remains a possibility.

GLOW-IN-THE-DARK NITS

To aid the removal of head lice from children's hair, Yale paediatrician Sydney Spiesel developed a shampoo containing a dye called blankophor that attaches to the chitin coating of the nits (louse eggs). When lit by ultraviolet light, the nits glow brightly, so that every last one can be easily spotted and removed, and the infestation ended.

PLEASURE BY REMOTE CONTROL

US surgeon Stuart Meloy patented a device that he first conceived in 1998 while treating a woman for chronic leg pain. He was implanting electrodes in her spine to provide pain relief when the woman 'suddenly let out something between a shriek and moan'. He asked her what the matter was, and she said, 'You're going to have to teach my husband to do that.' Meloy realized that he had placed the electrodes in the wrong location, and had just given the woman an orgasm. He went on to develop his device, comprising the stimulating electrodes, a small transmitter implanted under the skin of one buttock, and a hand-held remote control, which could give women orgasms whenever required. He published the results of his pilot study in 2006, having implanted the device in 11 women, who used it for nine days whenever they wished. All of those who used it (one said she was too stressed) reported that the device gave them pleasurable stimulation, including increased vaginal lubrication. Those who had previously lost their ability to experience orgasm recovered it, while those who had never experienced orgasm found it gave them pleasure, but did not send them 'over the edge'. One of the subjects of his study posed a tricky ethical question. 'Would it be considered adultery,' she asked Meloy, 'if I gave the remote control to someone other than my husband?' Meloy named the device after the orgasm-inducing machine in Woody Allen's 1973 film *Sleeper*: he called it the Orgasmatron.

SMARTER THAN YOUR AVERAGE BIRD

Researchers from Oxford University observed two New Caledonian crows bending a piece of wire into a hook shape to retrieve food by lifting a small bucket containing small pieces of pig heart, their favourite food, up a vertical pipe. New Caledonian crows are thus the only species apart from humans to make tools out of a material they do not come across in the wild. It was already known that these crows make a variety of tools in the form of bent, smoothed and plucked grass stems and twigs to obtain food, for example, by probing rotten trees for grubs. When one crow makes a refinement to a tool, he or she then shares it with others in the group.

Betty the New Caledonian crow retrieves the food-filled 'bucket' from the 'well' using a wire she has just bent into shape for this purpose.

Other species of crow also show great ingenuity. One group of crows in Japan has found a way to crack tough nuts, by dropping them onto a busy urban road at a pedestrian crossing. After a passing car has crushed the nut, they fly down to the curb to wait for the pedestrian light to turn green; they then walk out into the road and claim their nut.

CETOLOGICAL SCATOLOGY

Seeking to answer why the population of North Atlantic right whales had been reduced to only 350 individuals, a team of American scientists sought to find the answer in their faeces. From the whales' droppings, they hoped to find out what diseases, parasites or pollutants might be affecting the animals, and from looking at the hormones in the droppings, they could establish whether a whale was pregnant or nursing. The trouble was, it was very difficult to get hold of the noxiously smelly lumps of brown, orange or neon-red dung that the 50-tonne leviathans deposit at the surface prior to diving. The lumps quickly break up, and sink out of sight within an hour. Then Rosalind Rolland of the New England Aquarium had an idea. Zoologists in Washington state had for some years trained sniffer dogs to find the scat of grizzly bears, cougars, lynx and other animals. So why not adapt the training to whale dung? Before long, a black-and-tan mutt called Bob and a purebred Rottweiler called Fargo were recruited to Rolland's team, and found themselves poised in the bows of the research vessel, eagerly sniffing the waters ahead. When they located their quarry, the dogs would bark and wag their tails, and before long the retrieval of whale dung rose by a factor of four. Each successful retrieval was rewarded with a few minutes of play with a tennis ball. Once brought on board, the stinking dung was treated with extreme caution: 'If you spill it on your clothes,' Rolland explained, 'you want to throw those clothes away.'

STONED SPERMS' TIMING ALL TO POT

At the annual meeting of the American Society of Reproductive Medicine Dr Lani J. Burkman, assistant professor of gynaecology/obstetrics and urology at the University of New York at Buffalo, presented her findings regarding the effects of cannabis on male fertility. Burkman and her colleagues found that in marijuana smokers both the volume of seminal fluid and the total number of sperm were significantly less than in non-users. What is more, they found that sperm exposed to high levels of THC (the active ingredient in cannabis) were simply no good at their job. 'The sperm from marijuana smokers were moving too fast too early,' Burkman told the meeting. 'The timing was all wrong. These sperm will experience burnout before they reach the egg and would not be capable of fertilization.'

ROBOT'S FLESH-EATING TENDENCIES DENIED

The concept of an Energetically Autonomous Tactical Robot (Eatr) was pitched to the US military, and successfully secured funding from the Defence Advanced Research Projects Agency (DARPA). The robot, which could function as an ambulance or a gun turret and operate autonomously for months, is powered by a biomass engine, for which it gathers its own fuel. This led to rumours that the robot is designed to 'feed' on the corpses of dead soldiers, but this was strenuously denied in 2009 by one of the machine's inventors, Dr Robert Finkelstein of Robotic Technology Inc (RTI), who told Fox News: 'If it's not on the menu, it's not going to eat it.' The machine's sensors can apparently distinguish between animal, vegetable and mineral matter, and thus the robot should be capable of abiding by

Article 15 of the Geneva Conventions, which defines desecration of the dead as a war crime.

Offbeat ideas appeal to DARPA. In 2006 it was reported that it was considering a novel method of getting special forces personnel on to the roofs of buildings during terrorist crises. Each commando would sit in a chair that would be fired up a near-vertical ramp by a blast of compressed air (along the lines of an ejector seat). When the chair reached the top of the ramp it would be stopped dead, launching the occupant skyward. It was estimated that a person could be placed on top of a five-storey building in under two seconds. Firefighters and police forces would also benefit from the technology, DARPA said.

SHEEP AS AN ENERGY SOURCE

New Zealand farmers blocked the streets of Wellington to protest against their government's plan to introduce an annual 'fart tax' (also dubbed the 'back-door tax'). The gas produced by ovine and bovine flatulence and eructation is in fact methane, and as such is a significant contributor to global warming. It has been estimated that if properly harnessed the gas from New Zealand's 80 million sheep and cattle could supply all the energy needs of the country: the amount of methane produced by one sheep in one day could power a small lorry for a journey of 40 km (25 miles).

SIGNALS FROM OUTER SPACE?

(March) A radio source situated between the constellations Pisces and Aries, named SHGb02+14a, was observed three times, at a frequency

Energy the Cow on the steps of the New Zealand Parliament in Wellington during protests by farmers against the so-called 'fart tax'.

of approximately 1420 MHz. The discovery caused some excitement, as 1420 MHz is close to one of the principle frequencies at which hydrogen radiates and absorbs photons – and thus is thought to be the frequency that would be used by any advanced extraterrestrial civilization to transmit signals. However, there appears to be scepticism that this is the case in relation to SHGb02+14a.

FISH FRTS

An international team from Canada, Scotland, Denmark and Stockholm established that high-frequency underwater sounds were caused by streams of bubbles coming out of the anuses of herring. The noise, 'like a high-pitched raspberry', was characterized by the scientists as Fast Repetitive Tick (FRT). FRTs were only heard at night, and, as most fish other than herring cannot hear sounds at this high pitch, it was suggested that they might be a form of communication, enabling shoals of herring to keep together in the dark.

A PARTY PIECE

A 59-year-old Japanese mental-health councillor called Akira Haraguchi recited pi (*see* 1900 BC) to 83,431 decimal places. It is unlikely he was using the famous English-language mnemonic, in which the number of letters in each word represent the first 15 digits of pi:

> How I need a drink, alcoholic in nature, after the heavy lectures involving quantum mechanics!

THE AXIS OF EVIL

Kate Land and João Magueijo at Imperial College London found that hot and cold spots in the cosmic microwave background (the 'echo' of the Big Bang) were not randomly distributed through space, as would have been expected, but instead appeared to be aligned in a particular direction. Borrowing a notorious phrase from President George W. Bush, they dubbed this alignment 'the axis of evil', as it went against all predictions, and has remained a puzzle to scientists. It may be that the radiation is distorted into the observed alignment by the presence of a 'nearby' supercluster of galaxies.

LAZY RAYS DEFY EINSTEIN

The MAGIC gamma-ray telescope on La Palma in the Canary Islands found that high-energy rays from gamma-ray bursts in a galaxy 500 million light years away arrived four minutes after the lower-energy rays. This appeared to contradict Einstein's special theory of relativity, whereby light and all other forms of electromagnetic radiation, whatever their energy, travel at the same speed through a vacuum, i.e. 299,792 km (186,282 miles) per second. In 2008 NASA's Fermi gamma-ray space telescope picked up similar results from a source 12 billion light years away, and in this instance the time lag was 20 minutes. These observations raised hopes that it might be possible to test various theories that seek the Holy Grail of physics: the reconciliation of relativity and quantum theory via a quantum theory of gravity.

SCIENTIFIC AMERICAN APOLOGIZES FOR CHAMPIONING DARWIN

An editorial in the April edition of *Scientific American* expressed regret for the magazine's previous stance on a number of issues, including evolution:

> In retrospect, this magazine's coverage of so-called evolution has been hideously one-sided. For decades, we published articles in every issue that endorsed the ideas of Charles Darwin and his cronies. True, the theory of common descent through natural selection has been called the unifying concept for all of biology and one of the greatest scientific ideas of all time, but that was no excuse to be fanatics about it. Where were the answering articles presenting the powerful case for scientific creationism? Why were we so unwilling to suggest that dinosaurs lived 6000 years ago or that a

cataclysmic flood carved the Grand Canyon? Blame the scientists. They dazzled us with their fancy fossils, their radiocarbon dating and their tens of thousands of peer-reviewed journal articles. As editors, we had no business being persuaded by mountains of evidence.

The editorial was dated 1 April.

THE TIME TRAVELLER CONVENTION

(7 May) Current and former students at the Massachusetts Institute of Technology held a Time Traveller Convention on the campus. The event was (and continues to be) widely publicized in the hope of attracting some visitors from the future. Both the exact time (22:45 EDT) and the exact coordinates of the 'landing pad' (42.360007 degrees north, 71.087870 degrees west) were advertised, but as far as anyone is aware, nobody from the future turned up. The concept had one major flaw, in that given current physics, it is not possible to travel back in time further than the date on which the time machine first became functional.

EXPLODING TROUSERS EXPLAINED

In a detailed historical study, James Watson of Massey University, New Zealand, looked into various cases from the 1930s in which farmers' trousers had inexplicably exploded. It turned out that drops of the sodium chlorate weedkiller they were spraying on ragweed reacted with organic fibres in the trousers, which then spontaneously combusted, especially if exposed to heat or naked flame.

THE ARTIST WITH AN EAR ON HIS ARM

The Australian performance artist Stelarc (stage name of Stelios Arcadiou) had an artificial ear surgically constructed in his left forearm. At Edinburgh's International Science Festival in 2009 he announced plans to insert transmitters in the ear, so that people could hear what he is hearing. He also has plans to grow an earlobe from his own stem cells.

Stelarc at the Edinburgh Science Festival in 2009, showing the implant in his arm – an artificial ear made from a porous, biocompatible polyethylene material.

THE WORLD'S FIRST PENIS TRANSPLANT

Surgeons at Guangzhou General Hospital, China, successfully transplanted the penis of a brain-dead 23-year-old onto the 1 cm (0.4 inch) stump left to a married man of 44 who had lost the rest of his penis in an accident, and who was having difficulty urinating. Complex microsurgery was required to connect nerves and tiny blood vessels, and the procedure was regarded as a complete success from the technical point of view, with no signs of tissue rejection. But the introduction of the penis of a third person into a marriage can create difficulties. 'Because of a severe psychological problem of the recipient and his wife,' admitted Dr Weilie Hu of the Guangzhou surgical team, 'the transplanted penis regretfully had to be cut off.' Commenting on the case, Andrew George, a transplant expert at Imperial College, London, said 'Doing a penis transplant should be no more complex than anything else. But it takes time for nerve sensations to kick in and it's not clear whether the patient would ever be able to have sex with it.'

FINGERNAILS SCRAPING ON BLACKBOARDS

A team based at Northwestern University set up a range of experiments to discover why people hate the sound of fingernails scraping on a blackboard. They hypothesized that the sound might mimic the alarm call of a human ancestor, or the cry of a predator. In a separate experiment by Trevor Cox of the University of Salford, it was found that fingernails scraping on a blackboard came only in 16th place in a survey to find out which sounds people disliked the most. In the first three places were (1) the sound of someone vomiting, (2) microphone feedback, and (3) babies crying.

PRESIDENTIAL QUACKERY

President Yahya Jammeh of Gambia claimed that he had discovered herbal cures for HIV, asthma and high blood pressure, and proceeded to treat patients himself. Two years later Amnesty International reported that President Jammeh sanctioned a widespread witch hunt, in which as many as a thousand suspected 'witches' were kidnapped from their villages, stripped, beaten and forced to drink unknown hallucinogenic potions.

PRINCESS EXPRESSES SOLIDARITY WITH BABOONS

Researchers Dorothy Cheney and Robert Seyfarth published a book entitled *Baboon Metaphysics* indicating, among other things, that rank among female baboons is hereditary. When she visited them in Botswana, Princess Michael of Kent, daughter of an Austrian Nazi and wife of a cousin of the Queen, thanked them, saying: 'I always knew that when people who aren't like us claim that hereditary rank is not part of human nature, they must be wrong. Now you've given me evolutionary proof!'

BEAM ME UP, SCOTTY

(28 April) The small commercial *SpaceLoft XL* rocket blasted off from Spaceport America in Upham, New Mexico, carrying the cremated ashes of some 200 people, including those of Mercury astronaut Gordon Cooper and actor James Doohan, who had played Scotty in the early *Star Trek* TV series and films. Doohan had died in 2005, on 20 July, the anniversary of

the *Apollo 11* Moon landing. The rocket made a brief suborbital flight, entering outer space for four minutes before parachuting back to Earth, as planned. The following year, some of Doohan's ashes joined those of many others aboard another commercial rocket, *Falcon 1*, which launched from Kwajalein Atoll in the Pacific on 3 August, intent on making a low Earth orbit. However, the rocket failed after two minutes. The remainder of Doohan's ashes were scattered over Puget Sound in Washington state. *Falcon 1* has subsequently had more success, launching the RazakSAT satellite into low Earth orbit on 14 July 2009.

SEVENTY-YEAR-OLD GIVES BIRTH

(November) In Alewa, a village in northern India, 70-year-old Rajo Devi Lohan gave birth to her first child, Naveen. She and her husband Baba Ram had been married in 1950, but had failed to produce children. Naveen resulted from IVF of an egg donated by 'a good local girl', and then implanted in Rajo Devi, who, after birth, was able to breastfeed her baby.

DARK FLOW AND THE MONSTER OVER THE HORIZON

A team led by Sasha Kashlinsky from NASA's Goddard Space Flight Center found that a distant group of galaxy clusters are moving at enormously high speeds towards a small patch of space between the constellations of Centaurus and Vela. It has been suggested that this is most readily explained by the gravitational pull of some vast megastructure beyond the visible

Rajo Devi Lohan (70) with her daughter Naveen, whom she conceived via IVF treatment, and who was born by Caesarean section.

horizon of the universe. No such megastructures have been detected in the observable universe, so this would overthrow the received notion that the universe is pretty much the same throughout, and instead suggest that there is a 'tilt' in the universe, causing a mysterious 'dark flow'.

THE CRISPER CRISP ILLUSION

2008
Psychology

Charles Spence, professor of experimental psychology at Oxford University, concluded that people eating stale crisps rate them up to 15 per cent fresher if they are played tape recordings of crunching sounds: the louder the sound, or the more the higher frequencies are boosted, the fresher the crisps appear to be. Spence also collaborated with Heston Blumenthal, chef of the Fat Duck restaurant in Bray, Berkshire, where the eating of oysters was shown to be enhanced by the sounds of waves breaking, and that of bacon-and-egg ice cream by a recording of bacon sizzling in the pan.

MONKEYS TEACH OFFSPRING TO FLOSS

(March) It was reported that macaques living in temple ruins near Bangkok were teaching their young how to use fallen human hair to floss between their teeth, thus showing that other primates apart from humans are capable of teaching their children to use tools.

A TASTE OF RASPBERRIES AT THE CENTRE OF THE MILKY WAY

(April) Astronomers searching for signs of amino acids – the building blocks of proteins and hence of life – in the vast dust cloud Sagittarius B2 at the centre of our Galaxy, found instead ethyl formate, the chemical compound that gives raspberries their flavour.

PLEA FOR MORE LUNAR RESEARCH

(October) David Tredinnick, Tory MP for Bosworth, appeared to align his party with the astrological lobby when he spoke in Parliament: 'In 2001 I raised in the House the influence of the Moon, on the basis of the evidence then that at certain phases of the Moon there are more accidents. Surgeons will not operate because blood clotting is not effective and the police have to put more people on the street. I am arguing for more research.' The Royal College of Surgeons, the Royal Society for the Prevention of Accidents and the Association of Chief Police Officers all scratched their heads when asked to respond to Tredinnick's assertions.

BIRD AND BAGUETTE BLAMED FOR CERN FAILURE

(November) Testing of the Large Hadron Collider (LHD) at Cern, Switzerland, was interrupted when the power failed. Investigators found a bird eating a piece of bread at one of the locations where the mains electricity supply enters the collider, and this interruption caused a rise of temperature in part of the tunnel, disabling the LHD's superconducting magnets, which can only operate at 1.9°C above absolute zero. This was marginally less embarrassing than a 1996 incident involving CERN's Large Electron Positron Collider, which failed to do anything when it was switched back on after an upgrade. The problem, it turned out, was that workmen had left two empty beer bottles blocking the high-vacuum tube through which the beams were meant to travel.

THE LAST THING RABBITS NEED?

A team from the China Medical University Hospital, Taiwan, and the University Hospital in Zurich, Switzerland, successfully bioengineered rabbit penises. Male rabbits receiving the implants achieved erections and managed to impregnate females.

NOW YOU SEE IT, NOW YOU DON'T

(December) A motion with this heading was put before the Scottish Parliament by Gil Paterson, MSP. In full, the motion read:

> That the Parliament calls on the Scottish Environment Protection Agency, the nuclear regulator, to fully investigate the circumstances surrounding the reappearance of part of the 170 kg of enriched uranium that was first reported as being lost and then reported as an accountancy error in that the material was not missing but never existed in the first place; further calls for those in authority who are responsible for public announcements on decommissioning at Dounreay [nuclear power station] to stop spinning stories after the fact that infer that highly enriched uranium, which is not accounted for nor can be found, is in any event safe, and expects them to clean up not only nuclear waste but also their public announcements.

ON THE DANGERS OF BEING HIT ON THE HEAD BY A BOTTLE OF BEER

A team from the University of Bern in Switzerland studied whether it was more harmful to be hit over the head with a full bottle of beer, or an empty bottle. They concluded: 'Full bottles broke at 30 J impact energy, empty bottles at 40J. These breaking energies surpass the minimum fracture-threshold of the human neurocranium. Beer bottles may therefore fracture the human skull and therefore serve as dangerous instruments in a physical dispute.'

ALIENS COULD BE JUST LIKE US

(January) Simon Conway, professor of evolutionary palaeobiology at Cambridge University, told a conference at the Royal Society in London that any aliens visiting Earth are likely to have undergone a similar evolutionary process as humans, and ended up pretty much like us – greedy, violent, and with a tendency to exploit others' resources. 'Extraterrestrials won't be splodges of glue,' he said. 'They could be disturbingly like us, and that might not be a good thing – we don't have a great record.' They are less likely to be peaceful explorers, he said, than colonizers looking for a new place to live, or in search of water, minerals and fuel. Astronomers have now identified some 400 planets beyond our own Solar System, and some of these are in the 'Goldilocks zone' – where it is neither too cold nor too hot for water to exist in the liquid form essential for life. (Those of a whimsical bent have suggested the best place to search for a 'Goldilocks' planet would be between Ursa Major and Ursa Minor.)

HOW NIGH IS THE END OF THE WORLD?

Between the year 2169 and the end of the 22nd century, asteroid 1999 RQ36 has a 1 in 1400 chance of hitting the Earth, according to studies carried out by Andrea Milani and others at the University of Pisa. With a mean diameter of some 510 m (1673 ft), the asteroid could cause major devastation. However, an opportunity of averting such a disaster will present itself between 2060 and 2080, when the asteroid will make relatively close approaches to the Earth. If a rocket aimed at it achieved a deflection of only 1 km (0.62 mile), the chance of a future collision would be eliminated. The chances of another asteroid, 1950 DA, hitting the Earth on 16 March 2880, may be as high as 1 in 300. Given that 1950 DA is two to three times the size of 1999 RQ36, the result would be catastrophic. However, alternative calculations suggest that 1950 DA will miss the Earth by tens of millions of kilometres. If we escape these and other collisions – and manage to avoid destroying ourselves in other ways – the Earth will meet a more certain fate in 5 billion years time, when the Sun will follow the fate of other stars of its magnitude, and expand into a red giant. In the process, the Earth will be completely engulfed, and, as a consequence, vaporized. Life on our planet may have ended a long time before this point: it has been calculated that at the current rate of increase in solar luminescence, in just 1 billion years the Earth's surface will be too hot for water to exist in its liquid state, so ending life as we know it.

Index (of sorts)

cocaine injected into spine prevents subject feeling
  pubic hairs being pulled out 184
contraception
  using acacia and dates 16
  sneezing as an aid to 37
  contrary views as to efficacy of Coca Cola 283
crispy fibres, replenishing the 93

decapitation
  possibilities of blinking after 117
  role in love play of praying mantises 227–8
dedication to science beyond the call of duty
  death while inspecting volcanic eruption 35
  deliberately infecting oneself with venereal
    disease 102
  eating black vomit 134–5
  placing a noxious beetle in one's mouth 144
  losing consciousness at a great height 156–7
  swallowing a Petri dish full of bacteria 275
dogs
  driven mad by menstrual blood 42
  raised from dead, albeit temporarily 225–6
  hideous experiments involving puppies 248
Doppler effect, orchestra on train demonstrates 142
dreams
  of snakes inspire chemical breakthrough 160
  of dancing chess pieces disturb sleep of
    neurologist 226
ducks, plastic 284

Earth, the
  a mere 4000 years old 12
  shaped like a cylinder 18
  held to be flat 25–7
  circumference cunningly calculated 28
  fate of stone dropped down hole bored
    through 51
  lacks limbs, therefore cannot move 73
Ecstasy, on the origin of 199–200
eels, peculiar theories on the origin of 24
electricity, experiments involving
  sparks drawn from nose of 'Flying Boy' 92
  shocks administered to monks 94
  putting on and taking off stockings 99–100
  effects on the growth of turnips 103
  enlivening effect on dead frogs 115
  electrical rod applied to rectum of executed
    criminal 116
  electrode inserted into tooth socket 120
  ordinary people persuaded to inflict apparently
    near-fatal shocks 256

elephant
  consumption of boiled e. trunk 173
  expires after taking hallucinogen 257
end of the world 313–15
extraterrestrials
  intelligent, human-like creatures on the
    Moon 63, 137
  giant-headed creatures on the Sun 120
  communicating with Martians using
    mirror 193–4
  lunar insects 233
  formula for calculating possibility of 253–4
  quasar mistaken for 259
  pulsar mistaken for 262–3
  the Wow! signal 268
  a brief flurry of excitement regarding 300–1
  likely to be as unpleasant as humans 313

faeces
  applied to wounds to encourage formation of
    pus 52
  used to cure head lice 63
  weighing one's own 74
  highly valued 128
  echo of Big Bang mistaken for 260
  ejection of penguin f. 283
  development of synthetic f. 289
  moose indifferent to bear and wolf f. 290
  dogs detect whale f. 298
fish
  humans bursting out of 18
  fake 164
  furry 232
flatulence
  a prize for curing 111
  a device for absorbing 233
  in sheep and cows 300
  as a means of communication in herring 301–2
foot odour 281
frauds, fakes and hoaxes
  the boy with the golden tooth 63
  Linnaeus fooled by painted butterfly 88
  the curious case of Claude Émile Jean-Baptiste
    Litre 89
  animalcula blamed for pregnancies of otherwise
    chaste and virtuous women 96
  the chess-playing automaton 102–3
  a hoax that wasn't 122–3
  laws of physics apparently breached in
    Philadelphia and New York 129
  the transatlantic balloon express 141

Acknowledgments

The publishers would like to thank the following for permission to reproduce illustrations and photographs:

2 Science Photo Library; 6 The Bridgeman Art Library/Humboldt-Universität, Berlin, Germany; 10 Rex Features/Alinari; 13 The Bridgeman Art Library/Agnew's, London, UK/Private Collection; 21 Photo Scala Florence/Heritage Images/The British Library, © 2010; 30 Alamy/The Print Collector; 34 Photos12.com/ARJ; 37 Photo Scala Florence/BPK, Berlin, © 2010; 40 Corbis/Bettmann; 43 TopFoto/ The Granger Collection, New York; 54 Photo Scala Florence, © 1990; 57 TopFoto/Charles Walker; 64 Alamy/Interfoto; 67 Mary Evans Picture Library; 78 TopFoto/The Granger Collection, New York; 80 The Bridgeman Art Library/Archives Charmet; Bibliothèque de la Faculté de Médecine, Paris, France; 86 Mary Evans Picture Library; 91 Mary Evans Picture Library; 106 Alamy/Lebrecht Music and Arts Photo Library; 113 Getty Images/Hulton Archive; 124 Wellcome Library, London; 126 Photo Scala Florence/ White Images, © 2010; 134 Photo Scala Florence/ White Images, © 2010; 139 The Bridgeman Art Library/Royal College of Surgeons, London, UK; 146 Mary Evans Picture Library; 149 Alamy/Mary Evans Picture Library; 154 TopFoto/HIP; 161 TopFoto/ Charles Walker; 170 Corbis; 180 Getty Images/ Science & Society Picture Library; 186 Science Photo Library; 197 Getty Images/Popperfoto; 205 The Advertising Archive; 212 Corbis/Bettmann; 214 Corbis/Hulton-Deutsch Collection; 216 Alamy/19th era 2; 222 TopFoto/HIP/Oxford Science Archive; 229 Corbis/Bettmann; 238 Corbis/Bettmann; 243 National Library of Medicine, Bethesda, Maryland; 244 Getty Images/Keystone; 249 Corbis/Bettmann; 255 Corbis/Ross Ressmeyer; 268 North American AstroPhysical Observatory/ www.bigear.org; 273 TopFoto/Fortean/Blackmore; 278 NASA/Continental Dynamics Workshop/NSF; 287 Corbis/Science Faction/Louie Psihoyos; 297 Professor Alex Kacelnik/ Behavioural Ecology Research Group/Oxford University; 301 Getty Images/Fotopress/Ross Setford; 305 Rex Features/Photo by Capital Press Agency; 309 Getty Images/Barcroft Media/Barcroft India.

Quercus Publishing Plc
21 Bloomsbury Square
London
WC1A 2NS

First published in 2010

A catalogue record of this book is available from the British Library

UK and associated territories 978 1 84866 056 4
US and associated territories 978 1 84866 073 1

Printed and bound in China

10 9 8 7 6 5 4 3 2 1